What's Worth Aiming for in Education

Geraldine Mooney Simmie,
Manfred Lang (Eds.)

What's Worth Aiming for in Educational Innovation and Change?

Democratic mentoring as a deliberative
border crossing discourse for teacher education
in Austria, Czech Republic, Denmark,
Germany, Ireland and Spain

Waxmann 2012
Münster / New York / München / Berlin

Bibliographic information published by die Deutsche Nationalbibliothek
Die Deutsche Nationalbibliothek lists this publication in the
Deutsche Nationalbibliografie; detailed bibliographic data
are available in the internet at http://dnb.d-nb.de.

Findings from a European Comenius teacher education project 2006–2009
Gender, Innovation and Mentoring in Mathematics and Science (GIMMS)

Disclaimer: GIMMS was funded with support from the European Commission (grant 128749– CP – 1 – 2006 – 1 – IE– Comenius–C21). This publication reflects the views only of the authors, and the Commission cannot be held responsible for any use which may be made of the information contained therein.

  Education and Culture DG
Lifelong Learning Programme

ISBN 978-3-8309-2747-1

© Waxmann Verlag GmbH, 2012
Postfach 8603, 48046 Münster

www.waxmann.com
info@waxmann.com

Cover Design: Pleßmann Design, Ascheberg
GIMMS Logo: Michael Farren
Setting: Stoddart Satz- und Layoutservice, Münster
Print: Hubert & Co., Göttingen

Printed on age-resistant paper, acid-free as per ISO 9706

Printed in Germany

All rights reserved. No part of this publication may be reproduced, stored in a retrieval system or transmitted in any form or by any means, electronic, electrostatic, magnetic tape, mechanical, photocopying, recording or otherwise without permission in writing from the copyright holder.

Contents

What's Worth Aiming for in Educational Innovation and Change? 7

Foreword 9
Máire Geoghegan-Quinn, EU Commissioner for Research and Innovation

Acknowledgements 10

Part I: Educational Innovation, Change and Teacher Education

Chapter 1:
Democratic mentoring as a deliberative discourse for
educational innovation and teacher continuing education 15
Geraldine Mooney Simmie

Chapter 2:
What does it mean to be innovative in teacher continuing
education and mentoring? 25
Manfred Lang

Part II: National Country Reports

A) Developmental case studies

Chapter 3:
Ireland: reflective and collaborative mentoring as a deliberative discourse
with teachers, teacher educators and others 37
Geraldine Mooney Simmie and Sancha Power

Chapter 4:
Denmark: empowering innovation through experienced teacher – student
teacher democratic collaboration 52
Lotte Skinnebach and Birgitte Stougaard

Chapter 5:
Germany I: Subject-related dialogic and democratic mentoring in
biology teaching and teacher education 67
Doris Elster

Chapter 6:
Austria: video analysis for educational innovation,
mentoring and inclusion 80
Helga Stadler and Susanne Neumann

Chapter 7:
Spain: mentoring for innovative science education – teaching for achieving
scientific competence with student teachers and mentor teachers 89
Digna Couso and Roser Pintó

Chapter 8:
Czech Republic: constructivist approaches to innovation
in one school-university partnership 109
Eva Volná, Hashim Habiballa, and Rostislav Fojtík

B) Analytical case study

Chapter 9:
Germany II: case study design and results about innovation,
gender and mentoring 123
Manfred Lang and Gunnar Friege

Part III: Cross-Case Analysis, Findings and Implications

Chapter 10:
Educational innovation and change as deliberative discourse
across borders of educational systems in Europe: a cross-national
analysis of GIMMS findings 139
Geraldine Mooney Simmie and Manfred Lang

Bibliography 162

List of Authors 170

The Editors 171

Index 172

What's Worth Aiming for in Educational Innovation and Change?

Gender, Innovation and Mentoring in Mathematics and Science (GIMMS) was a European Comenius teacher education project, a school-university partnership across six European countries: Austria, Czech Republic, Denmark, Germany, Ireland and Spain between 2006 and 2009. GIMMS teacher educators and researchers engaged with teachers across their professional lifespan, in a philosophical and critical inquiry into teaching, innovation and mentoring. It involved science and mathematics teachers in lower secondary education.

At the time the argument for a newer type of teaching and teacher education was being raised by commentators from the social, political and economic worlds. Results from international achievement tests, such as PISA from 2000 to 2009, were raising questions about levels of literacy and numeracy. Policymakers were concerned with the strategic economic importance of this slice of the curriculum. Their concerns are addressed in the Foreword by the European Commissioner for Research and Innovation, Máire Geoghegan-Quinn and supported by Androulla Vassiliou, European Commissioner for Education and Culture.

GIMMS as a research and development study explored the pedagogical and political aspects of teaching and teacher continuing education. The cross-national team developed an evolving framework that regarded educational innovation and change as pedagogical and political text within the context of each nation state and across states (Aronowitz and Giroux, 1991; Giroux, 1988, 2009; Giroux and McLaren, 1986; McLaren, 1991a,b).

Innovative ideas and practices were realised in public spaces as a deliberative discourse with a diversity of actors including teachers at all stages of their professional lifespan, experienced teachers, student teachers and beginning teachers, teacher educators and policymakers. Overall the GIMMS approach was successful in generating ownership and innovation in teaching and teacher education. Part one positions the project within relevant literature. Part two presents the seven national case study reports from each of the six countries. Part three concludes with a cross-case analysis of findings and their policy implications.

GIMMS results make for compelling reading for policymakers, teacher educators, school leaders, administrators and teachers alike. Educational innovation and change clearly matter in the generation of 'new' pedagogical and professional practices for teachers, teacher educators and policymakers alike. Findings highlight the need for sustainable change of this nature to be underpinned by democratic mentoring relationships that extend well beyond institutional borders, be it school, university or state organisation. When fully supported, intellectually, morally, materially and structurally, as was the case in GIMMS 2006–2009, this type of deliberative discourse has the capacity for teachers to become agents of educational innovation and change in their classrooms and schools.

The GIMMS team
Digna Couso, Doris Elster, Gunnar Friege, Wolfgang Gräber, Manfred Lang, Rose M. Malone, Geraldine Mooney Simmie, Roser Pinto, Sancha Power, Lotte Skinnebach, Helga Stadler, Birgitte Stougaard and Eva Volna.

Foreword

Máire Geoghegan-Quinn
EU Commissioner for Research and Innovation

I accepted the invitation to write this foreword without any hesitation even though this pleasant task could have equally fallen to my colleague Androulla Vassiliou, the EU Commissioner for Education and Culture, since this publication deals with issues of common concern to us.

This book has a clear message: the need to bring about change in science and mathematics teaching practice, to help young people develop the key skills that Europe needs for smart economic growth and to make sure that we take care with *how we can motivate boys and girls to learn*. This message is also embedded in the Europe 2020 Strategy for smart, sustainable and inclusive growth that was adopted by the Member States earlier this year. This strategy will be delivered through a number of targeted initiatives such as the Commission's proposal for an "Innovation Union" and its proposals to support "Youth on the Move".

A common feature of both initiatives is the focus on improving key competencies in mathematics and science and improving general levels of educational attainment throughout the entire range from pre-school to tertiary level education. Yes, Europe needs more researchers, but it also needs more innovators and entrepreneurs, and people with creative and design skills. Europe will only prosper in the face of growing international competition from emerging economies if we move swiftly towards wealth creation through knowledge-based activities in all sectors of our economy. This will require commitment at all levels, national, regional and European. Specifically, this will require real commitment to change and to challenge the ways we currently do things in three key related areas: education, research and innovation.

This book highlights in particular the need for more gender awareness and gender support in how science and mathematics are taught in schools, and it sets out suggestions for effective mentoring approaches. I am sure that the experiences described will serve as important reference points for those of you who are able to bring about change, and I join with my colleague Androulla Vassiliou in commending this book to you.

Lifelong Learning Programme

Acknowledgements

GIMMS was funded with support from the European Commission (Grant Agreement number: 128749- CP – 1 – 2006 – 1 – IE– Comenius–C21). This publication reflects the views only of the authors, and the Commission cannot be held responsible for any use which may be made of the information contained therein.

We are indebted to national coordinators, teacher educators, researchers, teachers and schools in six European countries and a number of critical friends for their invaluable cooperation, critical thinking and contributions to the book. The GIMMS project team wants to thank the participants in each of the national case studies, whose experimental approach and insights into innovation and teacher professional learning through democratic mentoring offer valuable insights into the everyday reality of schools in a variety of contexts in six countries. These insights and findings form the basis for writing the book chapters that follow.

The national project team wishes to explicitly thank the participants and supporters of GIMMS in each of the participating countries. In Ireland these include Dr. Sancha Power, researcher to the project and a number of supporters of the project including Dr. Joanne Moles, Dr. George McClelland, Dr. Peter Childs, Tom Geary, Head of the Department of Education and Professional Studies, Mairead Condon, Administrator, the Chair of Education, Professor Marie Parker Jenkins and Professor Mary O'Sullivan, Dean of the Faculty of Education and Health Sciences at the University of Limerick.

GIMMS Ireland wants to say a special word of appreciation to the schools and teachers who took part and gave generously of their time and professionalism: Castletroy College, Co. Limerick, Villiers Secondary School, Limerick City, St. Caimin's Community School, Shannon, Co. Clare and Cóláiste Chiaráin in Croom, Co. Limerick. A number of people were also involved as external advisors to the project at various stages including George Porter and Niall Kelly, Inspectorate of the *Department of Education and Skills*, Anna Walsh Science Adviser with the *National Centre for Curriculum and Assessment*, Cathal de Paor, CPD Coordinator and Lecturer in Education in *Mary Immaculate College, Limerick City*, Moira Leydon, Education Officer with the *Association for Secondary Teachers of Ireland* and John Lucy, John Hennessy and Tim Regan from the *Second Level Support Services*.

The two German studies were kindly supported by the Institute for Science Education at the University of Kiel (IPN). Dr. Manfred Lang and Dr. Wolfgang Gräber from the IPN were appointed as national coordinators for the first half and the second half of the project respectively. The IPN is a research centre for science education with a nationwide function. The Institute's mission is to develop and promote science education through research. It is a member of the Leibniz-Association and is also affiliated to the University of Kiel (www.ipn.uni-kiel.de). Prof. Dr. Gunnar Friege was a member of the German project team, at the beginning supporting the cooperation with the Ernst-Barlach-Gymnasium in Kiel and other schools and later cooperating as a member of the University of Hannover.

GIMMS Germany I and Professor Dr. Doris Elster want to thank school partners who have supported the project. In Germany we want to thank the principals and teachers of the Gymnasium Marne and Gymnasium Heide, who have supported the piloting of the project. In Austria we especially want to thank the teachers who have supported us in the dissemination phase. Participating schools located in Vienna are the Gymnasium XIX Billrothstraße, the Gymnasium XIII Albertstraße, the Gymnasium XIII Wenzgasse, the Gymnasium XXII Strebersdorf and the Gymnasium Wiener Neustadt located in the federal state of Lower Austria. Within the University of Kiel the project was supported by the Head of the Department of Biology Education, Professor Dr. Horst Bayrhuber. At the University of Vienna the GIMMS project was supported by the Austrian Educational Competence Center Biology and Professor Dr. Schrittesser, Head of the Department of Educational Research.

A large number of student teachers, experienced teachers and beginning teachers were involved at various stages in the project. We want to thank them for their cooperation, their interest in the development of the practical course, and their readiness to take part in the evaluation process. GIMMS Austria wishes to acknowledge that the writing of their chapter would not have been possible without the splendid work of numerous student teachers. We would like to thank them for their reflective minds and great ideas which, without doubt, will progress science teaching in the future.

We would like to acknowledge support we received in the early phase of the project from Dr. Mary Billington and Cato Molte from Stavanger University in Norway. We also want to acknowledge the constructive feedback, critical thinking and guidance received from our external evaluator Dr. Rose Malone, Department of Education, National University of Ireland, Maynooth. We want to thank our internal evaluator Dr. Manfred Lang, Leibniz Institute for Science Education at the University of Kiel, Germany for his commitment to the project and willingness to share his vast experience of European projects.

Finally the project team is indebted to the European Commission for financial and project management support given to the GIMMS project 2006–2009. We also wish to acknowledge the corresponding support received from a number of higher education institutions including the University of Limerick, Ireland, University of Vienna, Austria, University of Ostrava, Czech Republic, Universitat Autònoma de Barcelona, Spain, University College Lillebaelt in Jelling, Denmark and the Leibniz Institute for Science Education at the University of Kiel, Germany.

Geraldine Mooney Simmie
GIMMS Project Coordinator
Faculty of Education and Health Sciences
University of Limerick, IRELAND
Geraldine.Mooney.Simmie@ul.ie

Part I:
Educational Innovation, Change and Teacher Education

> *"To encourage schools to become learning organisations requires ensuring among teachers: the motivation to create new professional knowledge; the opportunity to engage actively in innovation; the skills of testing the validity of innovations; and the mechanisms for transferring the validated innovations rapidly within their school and into other schools".* (Hargreaves, 2003, cited in OECD Teachers Matter, 2005, p. 130)

Hargreaves (2003) argues above that for teachers to lead their schools as successful learning communities they need to have the capacities to create, validate and transfer new professional knowledge and practices. This question is at the heart of the GIMMS project and the project argues that to do this successfully teachers need to become members of new sustainable networks that include other actors such as teacher educators and policymakers.

Innovation, teacher education, mentoring and gender were considered to be inextricably linked in developing progressive education. Over the lifetime of GIMMS 2006–2009 the team sought to develop a thread of discussion and debate to unify these diverse components and devise a framework for the study supported by an implementation plan.

National coordinators, mostly teacher educators and researchers, took part in a collaborative philosophical and critical inquiry into science and mathematics teaching and teacher education. The project aimed to bring about educational innovation and change with experienced teachers in science and mathematics classrooms and schools.

The GIMMS team perceived the role of the teacher educator in this change endeavour as complex and relational. The team needed both teachers and teacher educators to develop ownership of the project goals within a mutually enriching public *incubation space* that gave opportunity for a discourse about pedagogical issues. The GIMMS team sought to work in an organic and dynamic way with teachers and schools. This incubation space provided the context for the emergence of new ideas and practices that would assist the reconceptualization process. The positioning of the team and the project was supported by connection to relevant literature and the lived reality of everyday school experience.

Part I explores the arguments used in GIMMS as a development and research initiative. The key question explored how to develop educational innovation and change through a democratic platform of mentoring set within the classroom and school setting and with the capacity to reach well beyond the school setting to bring the most up-to-date research findings and knowledge to the professional work of teaching, learning and mentoring.

Chapter 1:
Democratic mentoring as a deliberative discourse for innovation and teacher continuing education

Geraldine Mooney Simmie
Faculty of Education and Health Sciences, University of Limerick, Ireland

> *"Teaching and teacher education are fundamentally political activities and that it is impossible to teach in ways that are not political and not value-laden ... part of teaching for social justice then, is deliberately claiming the role of educator as well as activist based on political consciousness and on an ideological commitment to diminishing (inequality)".* (Cochran-Smith, 1999, p. 117)

This chapter reviews relevant literature and introduces the evolving theoretical framework for GIMMS. Educational innovation and change in the project was conceptualised as pedagogical and political text, a deliberative discourse crossing borders within and between schools and other going beyond schools to other institutions (Giroux 1988, 2009; Aronowitz & Giroux, 1991; McLaren, 1991a,b; Pinar, Reynolds, Slattery, & Taubman, 2008). GIMMS sought to support and challenge teachers inside a web of democratic mentoring relationships of learning with colleagues, teacher educators and sometimes others, such as policymakers. It challenged traditional models of teaching and teacher education and gave teachers 'thinking time', 'direct access to research and up-to-date subject knowledge and research findings', 'structures', 'supports' and designated public *incubation spaces* to explore their teaching. It provided teacher educators with a unique cross-national opportunity to re-consider their own developing pedagogy in assisting teachers become agents of educational innovation and change (Loughran, 2006).

Research questions

Educational innovation and change, at the level of curriculum and teacher education, was interpreted by GIMMS as developing 'new' ways of teaching and teacher education through planning, reflecting and purposive debate with teachers, across their professional lifespan, about science and mathematics in lower secondary education.

Research indicated that subject matter knowledge needed to be taught in ways that were pupil centred, motivating and inclusive. Older conceptions of teaching positioned the science teacher as a deliverer of subject specific knowledge. Research findings were indicating the value of constructive feedback mechanisms and the intellectual, moral and social aspects of pedagogy. The GIMMS team adopted a sceptical positioning to any formulaic or technical rational approach to 'good teaching' or the development of innovation in the classroom or school. Clearly change and innovation was needed, how to get there was a debatable, ideological and political issue. School leavers entering a challenging and changing knowledge world required different capacities and dispositions than heretofore. For example, capacities such as critical thinking and problem solving were considered important. The role of the persona of the teacher and teacher educator in the endeavour of teaching, and in the work of teaching teachers how to teach, was being changed and challenged. These changes were not 'unproblematic' and required the development of a new and innovative discourse.

The deliberative discourse developed by GIMMS involved an extended network between teacher educators and teachers and other significant actors, such as policymakers. In each national case study teachers and teacher educators met in a variety of settings and groupings. The project was guided by a number of key questions including:
- What is needed to elicit motivation in learning science and mathematics and in motivating teachers to teach these subjects in pupil-centred ways?
- In what ways can educational innovation and change be introduced into the teaching and learning of science and mathematics?
- How can we develop new forms of collaboration and mentoring between teachers and between teachers and teacher educators, and significant others, for the improvement of teaching and teacher continuing education?

GIMMS interpreted the 'activity' of the project as involving strands that were intellectually, socially and culturally, and ethically supportive and challenging. This intellectual and relational approach moved teaching beyond the narrow confines of an individual reductionist top-down training model:

> "The teacher is an intellectual who generates knowledge, that teaching is a process of co-constructing knowledge and curriculum with students, and that the most promising ways of learning about teaching across the professional lifespan are based on inquiry within communities rather than training for individuals". (Cochran-Smith, 1999, p. 114)

Crossing Borders: GIMMS an evolving framework

The GIMMS team argued that teachers needed to network, together at their schools and with others beyond the borders of their classrooms and schools, in order to generate educational innovation and change. The team envisioned a framework for the project that captured this multiplicity of 'actors' and relationships (Figure 1). This web of interconnecting relationships included teachers, teacher educators and others such as policymakers.

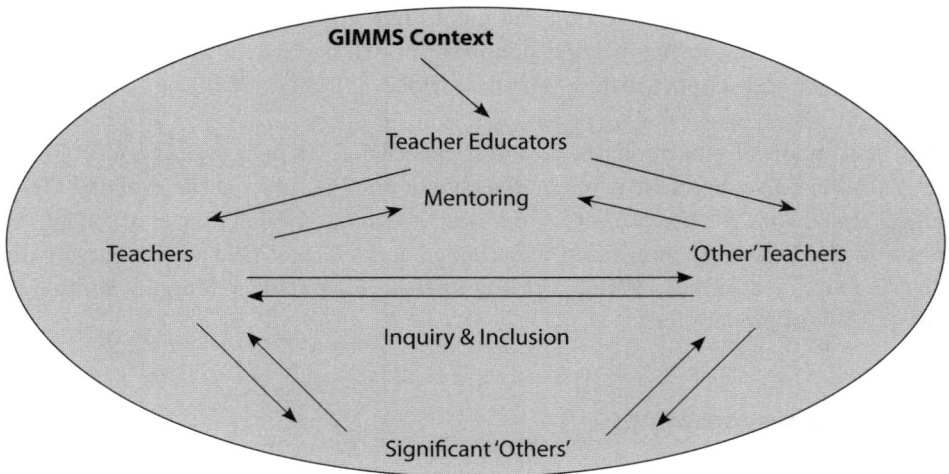

Figure 1: The GIMMS 2006–2009 collaborative framework

The first phase of the project involved finding interested school principals and science and mathematics teachers. This was followed by a period of negotiation and development where new and innovative approaches to pedagogical practices were developed, trialled and tested. In the dissemination phase protocols, products and practices generated were shared with a wider audience of teachers, schools and higher educational institutions through regional, national and international networking and conferences.

GIMMS as political text

GIMMS brought into focus the complex role of the teacher educator as an external agent of change in efforts to elicit educational innovation and change within the school community (Loughran, 2006). It raised questions about knowledge and power relations. For example, the difference in cultural and academic values between schools and higher education institutions. Schools and higher education institutions are not value-free and neutral sites but places where education is contested and politics and power struggles play out on a continuing basis.

Who is going to change in the school-university partnership process? Is it both teacher and teacher educator? The GIMMS team argued for the project to be mutually enriching and posited that the spirit of partnership was worth fighting for in the project. Teacher educators agreed to engage with critical reflection during the lifetime of the project to enter into their own change process as reflexive learners. The GIMMS team posited that educational innovation and change is 'relational' and involves diverse voices. These diverse voices needed, within an innovation process, to be liberating and not oppressive, to have equality of respect and to be based on a democratic framework of education and mentoring. Voice develops through a physical and intellectual journey beyond boundaries of classroom, of disciplines, of culture, of home and school learning (Shockley, Bond & Rollins, 2008).

This conception of educational innovation and change as pedagogical and political text, a deliberative discourse across a diversity of borders, became the evolving theoretical framework for the project (Aronowitz & Giroux, 1991, Pinar et al., 2008). It positioned educational innovation and change as a cultural field of discourse with diverse voices (McLaren, 1991a,b). This is further elaborated by Mooney Simmie & Lang (2012, in preparation).

Contextual responsiveness

The multiple threads within the overall tapestry of GIMMS – mentoring, innovation and gender – had different interpretations in each country. Crossley & Watson (2003) argued for an appreciation of diversity within comparative education studies. The cultural and contextual differences and commonalities across six countries added richness to the project and continually challenged the team's thinking. It generated *a creative dissonance* that became an important feature of the project and an important requirement for innovation within the public space. It was found that a cross-national comparative study has the capacity to bring teachers and teacher educators beyond local and national contexts and borders and introduce an international perspective, in this case a European perspective, to the matter under exploration.

Educational innovation and change matters

Petty (2009, p. 326) argues that the complex process of generating creativity and educational change requires that learners become willing to retain an array of possibilities and imagined practices rather than rushing quickly to forming uncritical solutions. A study by Dyer, Lindsay, Gregarson & Garber (2009) identified key attributes of innovative leaders, such as their capacities for *questioning, observing, experimentation, networking and association*. Their study suggested that innovators constantly question the unquestionable, observe how things function as they look for newer and better ways to do things, consistently experiment and learn from mistakes, proactively seek diverse perspectives through networking and finally bring all this information together through a process of association. They defined association as a form of thinking that connected seemingly unrelated ideas from different fields:

> "Associating, or the ability to successfully connect seemingly unrelated questions, problems or ideas from different fields, is central to the innovator's DNA." (Dyer et al. 2009, p. 63)

Science teaching and teacher education

Different schools of thought emerged in the 21st century to explain what was involved in science teaching. All placed different requirements on the persona of the science teacher in the classroom and school. One ideology, *science in context,* focused on teaching the concepts and principles of science through real life stories. These stories were drawn from real world and social settings but more often than not from industry (Bennett, 2003). Another ideological position involved teaching science through constructivist approaches. This required eliciting pupils own thinking about aspects of science and working with their lay theories to generate scientifically accurate new knowledge. This involved the teacher in a process of assisting pupils deconstructing old knowledge and reconstructing scientifically accurate knowledge (Driver, 1983).

Inquiry based science teaching was another recognisable movement in science teaching grounded theoretically and dating back to the work of Shulman (1987). He proposed that teachers have knowledge of teaching that enables them to cross the divide, and go beyond pedagogy and subject matter knowledge. He called this specialist knowledge base that teachers possess pedagogical content knowledge (PCK).

Asay & Orgill (2010, p. 58) argued that inquiry-oriented science teaching approaches are nonlinear and oftentimes messy. They are used in practice and go beyond a one-size fits all scientific method. They found that inquiry-oriented teaching requires components that engage the learner in: questioning that is scientifically oriented, giving priority to evidence, formulating explanations from evidence, connecting evidence to scientific knowledge and communicating and justifying explanations. They found that communication and explanation were among the least likely approaches to be used by science teachers in their classrooms:

> "Most in-service teachers who publish in The Science Teacher provide opportunities for students to collect and analyse evidence. Additional professional development experiences or examples may be needed to help teachers develop an understanding of inquiry that also includes the question, explain, connect and communicate features". (Asay & Orgill, 2010, p. 68)

GIMMS placed a strong emphasis on offering support for the development of communication among all stakeholders. It was particularly interested in developing the capacities of science teachers to publicly explain, justify, and re-conceptualise their professional practices. It did this through the development of public *incubation spaces* that promoted and supported both creative thinking and critical thinking about teaching and teacher education. Teachers developed their capacities in this regard through a deliberative discourse with peers, teacher educators and sometimes, with policymakers. In designing these public spaces, settings and groupings, GIMMS argued that teachers needed to be highly skilled and competent but that teaching, and indeed mentoring, involved much more than skill and competence (Mooney Simmie & Moles, 2011). GIMMS posited that creating this type of incubation space to engage with this work, the work of deliberative discourse, was essential for educational innovation and change to become a reality in the classroom and school:

> "After two decades of research on cognitive aspects of conceptual change, the questions of interest to science educators and teacher educators have shifted to: how do teachers and students create a space in which negotiation of scientific conceptions can take place? p. 49 … change in students' conceptions rests squarely under their own control but is intimately linked to the role of the teacher in facilitating and supporting discourse that can lead to change. Any instruction based on principles of conceptual change must assume the ability of a learner to reflect on the consequences of his or her thinking, to be metacognitive". (Beeth, 1998, p. 58)

Mathematics teaching and teacher education

Changes in mathematics teaching and teacher education mirrored changes that were taking place in the teaching of the sciences. GIMMS was guided by the research findings of the Dutch mathematics educator Freudenthal (1984), *real world mathematics,* and the work of Boaler (1997) and Jaworski (1994). This research highlighted the social nature of learning and the need for relevance and critical thinking in the teaching of mathematics. GIMMS mathematics teachers and teacher educators worked together as learning partners taking a collaborative, questioning and reflective stance.

Democratic mentoring as a deliberative discourse

Democratic mentoring relationships that crossed borders, and offered extended social and intellectual networks helped develop the capacity for educational innovation and change. Building these types of relationships of learning was perceived by GIMMS as problematic. Mullen (2005) noted that literature was divided between mentoring for socialisation and mentoring for change. Mentoring for socialisation required a top-down linear model of training while mentoring for educational innovation and change was going to need a more sophisticated, nuanced and organic approach. Mentoring was conceptualised in the project as democratic, reflective and collaborative. Mullen (2005, p. 73) argued for an interdependent stance for all actors in such partnerships:

> "The co-mentoring or collaborative structure of learning focuses on mutuality and the value of interdependent, reciprocal learning that challenges assumptions about hierarchy, rank, status – and, consequently, who is 'teaching' and who is 'learning'."

The GIMMS team interrogated a number of conceptions of mentoring including those advanced by Maynard & Furlong (1995). They presented a three stage framing moving from skills and competence training through an apprenticeship of observation to a final stage involving critical co-inquiry. While none of the stages were mutually exclusive they assisted in describing different interpretations of mentoring witnessed and developed in each of the national case study reports. The GIMMS team posited that democratic mentoring for educational innovation and change required a critical co-inquiry model that crossed borders and was underpinned by egalitarian theory (Lynch & Lodge, 2002, Mooney Simmie & Moles, 2012). This conception of mentoring, as will be seen in Part II, strongly challenged the dominant discourse of teachers and schools across each of the six countries.

Gender awareness

In the 2000s statistical data indicated that many girls were opting out of continuing study in the physical sciences and mathematics, especially at the more theoretical level, thus denying them access to significant career opportunities:

> "TIMSS data for European countries indicate that gender differences in favour of males appear at the fourth grade and continued through to the final year of secondary school, with the gender gap widening at each year (IEA 2000) … by the eight grade, students' attitudes toward science and mathematics begin to show gender differences. More males than females believe it is important to do well in science and mathematics to get their desired job. These differences may account, in part, for the low participation rates of women in higher levels of science in Europe". (Scantlebury & Baker, 2007, p. 268)

Curriculum as gender text first entered educational discourse in the 1970s coinciding with the women's liberation movement and international concern for equality of opportunity between boys and girls. Gender was, at that time, particularly focused on participation rates and access possibilities to these subjects at all levels. Throughout the 1980s and 1900s feminist researchers challenged the nature of the curriculum, pedagogical styles, classroom interaction and other practices that continued inequity between boys and girls. By the 2000s gender was no longer viewed as fixed and a process model was advanced:

"The emphasis on process shifts attention to exploration of how meaning is constantly being reproduced and negotiated, and can have unexpected and contradictory effects. This provides a framework for understanding social change and the way in which individuals, through this process of negotiation with meaning, are constituting their world". (Alsop, Fitzsimons & Lennon, 2002, p. 80)

GIMMS teacher educators shared diverse perspectives on gender. Fear of stereotyping girls was mentioned by some whilst others posited that gender differences in science and mathematics were not significant at lower secondary education. The focus was on raising awareness of gender with teachers as a social justice issue (Cochran-Smith, 1999, 2005).

GIMMS guiding principles

The theoretical framework for GIMMS emerged over time shaped by the selection of relevant literature and national policies and practices in each of the six countries. It resulted in the development of shared principles which helped interweave various threads of the project:

Principle 1: The team situated teaching within a continuum and professional lifespan with knowledge and insights flowing in non-linear and dynamic ways between student teachers, beginning teachers, experienced teachers, teacher educators and other significant actors, such as policymakers.
Principle 2: The team asserted that teacher educators needed to find innovative, inclusive and critically engaging ways to work alongside teachers, ways that were democratic, and displayed equality of respect, across the teacher's professional lifespan.
Principle 3: The team posited that changing pedagogical practices to inquiry-oriented, inclusive and communicative learning involved a deep cultural challenge and conceptual change for teachers, teacher educators and policymakers as well as a change in practice.

GIMMS national case studies

GIMMS followed a case study approach and was methodologically guided by principles advocated by theorists such as Hammersley (1986), Merriam (1998) and Yin (2003). Table 1 shows the processes and final products developed in each of the case studies.

Table 1: GIMMS 2006–2009 processes and products.

Country	Processes	Product(s)
Austria	Video-analysis for teacher learning egalitarian, democratic and organic models of mentoring between teacher educators and teachers.	Key questions for analysing physics teaching, including gender awareness questions. Reflective account of work with student teachers.
Czech Republic	New ways of working with pupils in groups that were social and motivating. Largely novice-expert relations.	CD of a range of interactive classroom materials for teaching physics, mathematics and computer science.
Denmark	Models of egalitarian mentoring that included face-to-face meetings, seminars and a virtual learning platform for teaching as reflective, inclusive and research practice.	Classroom resource material in Nuclear Energy, including web-pages and videos, from a multiplicity of perspectives including gender awareness.
Germany	Project I used a comparative study of mentoring in different states Project II developed an innovative approach to teaching biology taking ethics into account.	A web-page for the project and a range of reflective and innovative classroom resources on the teaching of human biology taking ethics and gender inclusion into account.
Ireland	Using democratic and reflective models of mentoring to develop higher order co-operative thinking among science and mathematics teachers.	A resource pack with a multiplicity of approaches to teaching aspects of chemistry, physics, biology and mathematics using inquiry-oriented and reflective approaches.
Spain	Developing an interactive model of mentoring between student teachers, teacher educators and experienced teachers for co-learning and co-ownership. Largely novice-expert mentoring relations.	A resource pack, and web-site pages, for innovative approaches to planning, teaching and reflecting on materials science
GIMMS General	Initial web-site www.ul.ie/gimms	Final web-site www.gimms.eu/

Educational innovation and change as a deliberative discourse that crosses borders

In conclusion, the GIMMS project, conceptualised educational innovation and change, as requiring a public space – an *incubation space* – for deliberative discourse that crossed borders between teachers, teacher educators and significant others, such as policymakers. This discourse was underpinned by key ethical principles and implemented on a platform of democratic mentoring that sought to inquire into 'new' ways of teaching science and mathematics while taking social justice issues, such as gender awareness, into full account. It challenged traditional models of teaching and teacher education. GIMMS argued for the need for teachers and teacher educators to be continually involved in public and reflexive ways to explain, justify and re-conceptualise their professional practices.

Nixon (2004, p. 119) defines deliberation as a form of thinking-not-yet-finished which seeks to argue beyond the point of seemingly ultimate disagreement. This involves continually questioning, taking time to 'stop and think' and is connected to what Socrates called 'negative wisdom', a sense of one's own relative ignorance. In the case of GIMMS this negative wisdom was about what constitutes 'good teaching', 'best practice' and 'good teacher education'. GIMMS positioned educational innovation and change as work that honoured the intellectual nature of teaching and required deep social and critical engagement between teachers and teacher educators and a number of other actors (Giroux, 1988).

Deliberative discourse within a higher education institute and university setting is most often associated with 'rational argument and public oratory' (Nixon, 2004, p. 121). Within an extended social democratic network, such as a school-university partnership, this clearly needed to be redefined and reconceptualised. GIMMS upheld this pedagogy of recognition when designing its deliberative border crossing discourse framework for educational innovation and change:

> "The authenticity of the messy narrative (the story, the autobiography, the biography, the anecdote etc) needs still to be upheld against the presiding influence of the tidy syllogism. What is at stake is not the survival of a particular form of public language, but my capacity for expression and receptivity in respect of diverse forms of communication. Pedagogies of recognition acknowledge these communicative and expressive differences." (Nixon, 2004, p. 121)

Chapter 2:
What does it mean to be innovative in teacher continuing education and mentoring?

Manfred Lang
IPN, Leibniz Institute for Science and Mathematics Education, Kiel, Germany

In GIMMS collaborative models of mentoring were used to introduce and study educational innovation and change in science teaching and teacher continuing education. Collaboration was mostly realized in small scale studies as continuous mentor-mentee dyads, but also in collaborative groupings of mentors, other experienced teachers, teacher educators and groups of mentees. For the development of a collaborative culture of innovation these settings need to be institutionalized for a continuous development of innovation. Supporting a regular incubation setting, that crosses borders, for this type of extended collaborative exchange will be a challenge for schools, for higher education institutes and for policymakers alike.

In this section we take a closer look at the national case studies of GIMMS to find out how collaboration in teacher education might work in different educational systems for the realisation and institutionalization of intended innovations. Concerning teacher education a special emphasis is given to the aspect of collaborative and reflective mentoring. Curriculum innovation focuses on the generation of a new culture of pedagogical and professional practices, with gender sensitivities, within a framework of teacher continuing education and mentoring. For the development of our viewpoint about collaborative teacher education and innovation we mainly analyzed country reports and interviews, apart from direct observations in some schools and discussions with teachers and teacher educators.

Let us start with the description of basic structures of the initiative, which are necessary in order to trace and understand outcomes about the dynamic and the kind of educational innovation and change developed in science teaching and teacher education.

Background and settings of the national GIMMS studies

In *Ireland* the case study schools and teachers agreed to work on a number of aspects of the junior cycle curriculum. Junior cycle syllabuses in science and mathematics had only recently been revised in lower secondary education system. The changes introduced mandatory science experiments and pupil-centred approaches to teaching and learning science and mathematics. Three schools were working on aspects of science innovation – data logging in science education, ICT-enhanced learning and eco-friendly science – with a fourth school developing their teaching of mathematics. Innovation in all schools was interpreted in GIMMS Ireland as teachers working together, using higher-order collaborative discourse and practices that not only included discourse within their school settings but also discourse beyond their school settings, with teacher educators and others. This approach was aimed to build capacity to develop reflective teaching and mentoring approaches to professional learning, for teacher educators, teachers and their pupils. Teacher continuing education in the project was based on generating a discourse of co-inquiry and reflection, using a platform of democratic mentoring, between teachers and teacher educators, in one school-university partnership. This deliberative discourse sometimes involved policymakers.

In the *Czech Republic* four case study schools with eight teachers got involved in the project. The focus was on a change to constructivist approaches to teaching physics, mathematics and computer science. Topics chosen were number, weight, time, geometry, conic sections and transformations of the plane. Teaching is usually done with teachers active and pupils passive. This needed to be changed. The project explored the concept of the teacher as mentor and connections between pupils' own ideas and scientific thinking. Constructivism as a teaching approach aimed to develop viable mental models for pupils' scientifically accurate thinking processes. Teachers worked with teacher educators to design lessons. Back in the classroom the teachers conducted the lesson taking care to include pupil discussions and to give some 'space' for pupils to explore topics.

In *Spain* four schools were participating in the materials science project, a curriculum innovation in a separate European project. Spain has had a big push to develop science education after a number of 'below average' PISA outcomes. The case study involved examining mentoring models between experienced teachers, student teachers and teacher educators. Teachers focused on learning about scientific competencies in context-oriented science education. Teachers perceive themselves as mainly qualified as subject matter knowledge specialists. In mentoring a very strong novice-expert system exists. Mentees spend most of their time sitting in on lessons given by a mentor, who in turn is not given any additional professional support or continuing education for this role. The GIMMS study gave an opportunity to explore a new mentoring relationship of learning. Gender differences in science did not strongly feature as it was believed that these differences did not appear so much in this age group as they did at a later age. In addition it was contested that research findings showed that taking context into account, as GIMMS *Spain* did, would advantage both girls and boys and make it less necessary for a specific focus on gender.

Project 1 in *Germany* involved the design, implementation and evaluation of a cross-curricular course linking various topics in biology to the study of number of ethical dilemmas. This course was designed for biology student teachers. New innovative gender sensitive teaching materials were developed and reflected upon, and this was accompanied by collaborative, democratic and dialogical mentoring with student teachers on a number of different levels. The aims were to link curriculum change and teacher continuing education through piloting a number of mentoring models and through the identification, development and dissemination of innovative gender proofed practices in biology teaching. The case study started with two mentoring teams including fourteen student teachers, one science educator and two experienced teachers, acting as mentor teachers in the federal state of Schleswig-Holstein. Mentors and mentees wrote reflective journals, using photographs as border crossing tools for reflection on classroom experiences. In a second stage dissemination was completed in Vienna, Austria with six mentoring teams consisting of one hundred and fifty two student teachers, one science educator as mentor and six experienced teachers.

Project II in *Germany* made an assessment and analysis of two different types of teacher professional development as the second phase of a more practically oriented preparatory teacher education programme. Research was conducted in two different federal states, one promoting a new model of mentoring with modules offered by a state institute and another fostering more traditional methods. Mentoring in this instance supported initial teachers' learning process and practical development in the school setting. A central question about innovation in this field of relationships was how to develop better partnerships between student teachers and experienced teachers for continuing teacher professional learning. Teacher education and mentoring in school environments are seen as central efforts in building innovative processes as a consequence from the *Programme for International Student Assessment* (PISA).

In *Austria* the Ministry was starting to introduce national testing and beginning to centralize the education system. In addition it was seeking to involve more girls in science and technology. In *GIMMS Austria* a structured and reflective approach for physics teaching was designed for student teachers already participating in the student teaching phase of their education. The main aim was to support them by developing and discussing lesson concepts through personal forms of mentoring that engaged student teachers in a reflective process about their own teaching methods. Topics included pupils' ideas, gender issues, teaching methods and innovative approaches such as e-learning, science lab or the role of experiments. Student teachers were introduced to a reflective mentoring approach as reflective practitioners in order to prepare them to interrogate and research their own teaching. Teacher educators used videos and action research cycles with student teachers to critically inquire into issues in teaching, with a specific focus on gender issues in the teaching of physics. Learning was constructed, between the teacher educator as mentor and the student teacher as mentee, as a democratic relationship of mentoring.

In GIMMS *Denmark* two case study schools focused on experienced mentor teachers working with student teachers while another two schools explored better ways of developing science and mathematics teaching through experienced teachers' cooperative interactions. Pupils' needs were at the centre of the work with teacher educators, student teachers and experienced teachers. The focus on curriculum innovation was to raise pupil motivation and gender awareness practices, as equal opportunities for boys and girls, through a model of innovative mentoring focusing on collaborative learning, effective group work and selected materials for gender sensitive teaching. Experienced teachers, student teachers and teacher educators worked together to develop teaching which reflects research findings of gender-sensitivity. Teachers, at the lower secondary level, generally focus on thinking about pupils' learning needs. Subject matter knowledge had less pride of place in the Danish education system at this level at this time. Often teachers are not qualified graduates in these subject areas. Educational innovation and change was interpreted in GIMMS Denmark as a change in the organization of the learning environment in the classroom and school and a border crossing approach to innovation in teacher education. For example, in the classroom change focused on a variety of eclectic pedagogical changes other than whole class teaching. It meant having new groupings, new materials, using inductive teaching approaches and connecting practice elements to theory. Mentoring relationships of learning between student teachers, experienced teachers and teacher educators were initiated through a seminar at the beginning of the project. At this time there was no tradition of mentoring for change and innovation in Denmark. In the GIMMS project experienced teachers and student teachers were starting to collaboratively work together in lesson planning and doing this with the support and challenge of teacher educators. In this way innovative pedagogical and professional practices were being developed and explored.

Insights about the dynamic of innovation in teaching and teacher education

The preceding outlines of seven case studies from Ireland, Spain, Germany, Czech Republic, Denmark and Austria demonstrate a diversity of presuppositions in educational systems and project assumptions and expectations about innovation, mentoring and teacher continuing education. In chapter 10 we use final interviews of national partners, conducted in 2009 at the end of the GIMMS study, and other data sources to trace these assumptions and expectations and develop a deeper understanding of the process of innovation. Not all expectations from research or policy requirements in the national projects could be realized in school practices for different reasons and in some cases new unexpected results and outcomes were found. In the final interviews we might find some insights into this dynamic on educational innovation and change in science teaching and teacher continuing education and mentoring.

In the interviews national coordinators were asked core questions about their national policy with regard to science and mathematics education, pedagogical practices in class, types of innovative practices and resources in the teaching and learning of science and mathematics, gender awareness, models of mentoring, teachers' change of practices as a result of GIMMS, kinds of cooperation and reflective communication. In addition some questions about the context of the study were also asked. For example, a self-rating about the coordinator's role in the study and the capacities of the teachers involved. Some of these questions may get us closer to a better understanding of the innovative process, the multiplicity of roles required, and the multiplicity of actors required, such as teachers, across their professional lifespan, teacher educators and additional stakeholders, including policymakers.

Educational innovation and change

In all national case studies educational innovation and change – at the level of curriculum and at the level of teacher continuing education – was judged by the coordinators to be central. During the lifetime of the project, over three years from 2006–2009, it was successfully realised.

GIMMS *Ireland* led to profound changes in practice among the teachers involved. Teachers developed innovative protocols, products and practices which were trialled and tested in classrooms and found to make the teaching of science and mathematics more motivating for pupils and teachers alike. The teachers began to work in higher-order professional collaborations together – through discussing teaching, learning and innovation and engaging in a number of reflective practices including writing reflections. This reflective writing process was conducted alongside teacher educators, with the project team of teachers, and with their pupils. It was guided through a deliberative border crossing discourse that included the teachers, teacher educator/researcher and others, including policymakers. The intrinsic value of reflective writing for deepening and improving thinking skills was more fully realised by teachers when they involved their pupils in the process. This resulted in a 'learning partnership' been established between teacher educators, teachers and their pupils. This reflective stance to innovation was a timely initiative inside schools given the levels of business reported within schools in Ireland, the review of the state examination system in compulsory education and the national policy impetus for improving literacy and numeracy among teachers and pupils alike.

A central innovative outcome of the *Czech Republic* project was that pupils positively evaluated and enjoyed constructivist lessons using information technologies and less traditional teaching. Usually there are detailed lesson plans to be followed and all topics and activities documented in a class book controlled by inspectors. In the GIMMS project a constructivist approach allowed pupils more freedom to discuss their ideas and work in groups. However, teachers in the project were not empowered to change much because of strong government controls, not enough time for collaboration and no additional professional support for mentoring as an agent for educational innovation and change.

The *Spanish* project found that their teachers showed some interest in innovation, although in general teachers adhere to traditional subject matter knowledge and teaching materials in line with the official syllabus. Different from this long-standing tradition the teachers were interested in new ways of teaching and mentoring and achieved some understanding of a learning community of teachers and teacher educators. But mentoring as an innovative source in school needs to be developed and evaluated. There is no continuing collaborative education for teachers as mentors. Good teachers are automatically considered to be good mentors. Therefore there was a need to develop specific mentoring activities in the project and to analyze these activities: "The main mentoring research was about new ideas associated with innovation. But when the idea of innovation arises, we don't know how to measure the quality of activities. How can we discuss activities whether they are better or worse?"

In *Project 1 in Germany* the reflective and democratic model of mentoring was judged to be a successful promising way to change mentoring in teacher education from a top down expert-novice model to a model characterized by flat organisation, mutual respect and dialogic exchange that crossed borders between schools and university. This new approach was applied in different contexts in pre-service and in-service teacher education and showed innovative ways of developing a border crossing partnership approach to educational innovation and change. The latter was done by fostering extended professional networks, inquiry and reflection between student teachers and mentors at the school site and at the universities.

In *Project II in Germany* student teachers agreed that an innovation with regularly changing contact personnel, a number of different actors in teacher education supported the attainment of professional autonomy and critical feedback for student teachers to a certain extent. It also produces a certain amount of confusion for student teachers as some felt the need to follow the guidance of different masters in school and state institutes. School-based teacher education, it has been suggested, can facilitate decisions for examination of student teachers and reduce boundaries between subjects as suggested in the following coordinator interview:

> "What I consider as innovative in Schleswig-Holstein is the school-based examination lesson and that this lesson is not dependent on subjects. It is possible to discuss such a lesson of History or English or Physics within the school. But it is beyond disciplines; they are going towards other traditions in other subjects such as pictures on the board or experimenting. This is from my view very innovative, but with the consequence that much of the teacher education is given to the schools". (National Coordinator, September 2009)

The *Austrian* course about teaching physics helped student teachers to reflect and complete action research assignments on their own teaching methods about different topics as "reflective practitioners". They got involved in innovative topics such as science labs or ICT. This was different from student expectations that are mainly focused on subject matter knowledge only and was much appreciated. It encouraged students to learn more about teacher education itself, mentoring and appropriate use of teaching methods. Mentoring was understood as a new way of dialogical co-inquiry and reflection in an extended public forum including student teachers, experienced teachers and teacher educators.

In the *Danish* project many of the innovative issues such as gender awareness, democratic mentoring, student motivation and new teaching organization were also found in the official national policy documents and discussion. Gender was not seen by the experienced teachers as an important topic. Awareness in this topic was developed by the teacher educators through readings from Sjöberg (2007a,b) and others and findings from the OECD PISA 2000, 2003, 2006 studies (OECD, 2001, 2003, 2004, 2007) . Mentoring changed during the lifetime of the project from traditional supervision to a more collaborative model where student teachers prepared lessons together with experienced mentor teachers and teacher educators. Different types of classroom organization were explored. This had implications for educational innovation and changes in subject matter knowledge and teaching and learning approaches. Subjects were generally better related to context and teachers used more sophisticated methods and materials than traditional transmission of subject content.

Findings about successful teacher education and mentoring in GIMMS *Denmark* were disseminated in a variety of ways. For example, results were disseminated on a virtual platform to all experienced teachers, to a number of different education boards and were used to teach student teachers at the university college and through teacher in-service courses. However experienced teachers were not generally well prepared for innovation and change:

> "Teachers are on their way. We are moving in the right direction to a new thinking about school. We expect from teachers more professionalism. That is a new issue. Teachers who were educated during the last ten years are prepared to it. But the teachers before this period are not prepared". (National Coordinator, September 2009)

Concluding comments

Gender got special attention through PISA results, which demonstrated considerable differences in science and mathematics achievements for boys and girl in many countries. However, in most cases of GIMMS gender was not interpreted as the sole topic of inclusion. There was resistance to engaging with this topic and a felt resistance to giving gender a critical gaze by science and mathematics teachers. It was interpreted somewhat differently in all case studies. For some it was part of a bigger multicultural problem for others it was a question of equal opportunities for male and female pupils or student teachers. Gender as an issue of importance in the sciences and mathematics clearly needs further intervention, development and research across Europe at the level of lower secondary education.

The equality issue that was successful in the project was the setting up of mentoring as a democratic platform between a variety of actors, teachers across their professional lifespan and teacher educators, and on some occasions policymakers, in ways that levelled the opportunities for mutual learning and enrichment.

In spite of all the differences in the background and dynamic of innovation and teacher continuing education in different national systems, there was one consistent driving factor in all case studies: the interaction of teachers across their professional lifespan with a university or research and development higher education institution. This answers the question of the first section, how teachers become actors in educational and curriculum innovation: As actors they need collaborative partnerships to cross borders from the narrower confines of the classroom and school practice and the world of research and academia. Access to the touchstone of academia provides a valuable lens into changing pedagogical and professional practices and retains, for experienced teachers in particular, a connection to the specialist knowledge base of teaching. A reduced collaborative partnership within a school, as a school-based teacher initiative only, which was proposed in the German project II, while necessary, might not be sufficient to generate educational innovation and change. While teacher-teacher discourse might give schools more autonomy and start to build collaborative networks of teachers it would not necessarily give the additional impulses from other external key actors to become involved in the critical thinking and creative orientation necessary for sustained educational innovation and change.

GIMMS interactions established a stable structure for exchange, generation of new ideas and sources for innovation and teacher continuing education. In all case study reports cooperation with teachers and teacher educators triggered discussions about different ways of teaching, mentoring or gender awareness as innovation within curricular frameworks. This would not have happened in isolated school cultures. Specific details of the full range of activities and conceptual change under discussion in each national case study in dealt with in greater detail in Part II of this book.

Cooperation and collaboration across borders, as a driving force for educational innovation and change, was not necessarily disseminating the new ideas and conceptual changes system-wide into other schools nor was it totally sustainable beyond the lifetime of GIMMS 2006–2009. The national activities in GIMMS were local initiatives with additional resources funded by a number of policymakers, the EU Comenius project and teacher education institutions. This level of funding provided the designated 'thinking time' and 'meeting time' required for the various actors in the project to meet and exchange their thinking and research findings. Funding was also provided for a number of 'supports' to engage in this type of work, both 'intellectual and moral supports' such as guiding principles and quality time and 'material and structural supports' such as financial and administrative resources, meeting times, minutes of meetings so that protocols, products and practices developed could be disseminated to other teachers, teacher educators and policymakers. While each of these involved a modest investment in personnel, time and finances this dynamic of innovation and equality of respect could not have taken place without these additional practical support mechanisms.

However, there were some perspectives of institutionalisation that might carry on with these first innovative steps. For example, institutionalisation is obvious in Denmark with the nomination of a coordinator onto a board of education. It is obvious in Ireland with a policy curriculum workshop to debate issues with local and national policymakers and teachers alike (Mooney Simmie & Power, 2012). In each case an interest in conceptual change, contextual change and practical change is signalled and installed in national policy. It is this political relation that, through connection to national policymakers, gives these initiated school improvements and innovations a better chance to influence policy and become a sustainable development for innovative science teaching and the continuing education of science teachers in a knowledge society.

Part II: National Country Reports

GIMMS 2006–2009 recognised the different contextual landscapes in each country and each professional setting. These contextual differences were found to be invaluable in exploring and understanding educational innovation and change as a cultural and political dynamic.

This section presents the seven national case studies, from each of the six countries. Each author seeks to give the reader a glimpse into each national policy process and charts the variety of context-specific local initiatives taken within each case study to develop a discourse on educational innovation and change between teachers, teacher educators and with, in some instances, other key stakeholders, such as policymakers.

The case study reports explore the real world experiences of GIMMS teachers, at various stages in their professional lifespan, and in a number of different groupings and settings as they interacted with national coordinators, teacher educators and researchers to produce innovation within their classroom and school. They present the reality of the day-to-day struggle of the classroom and school. In all cases new and innovative approaches were generated that were found to be motivating and enjoyable for both pupils and teachers alike.

The case studies highlight the possibility and problems involved in opening a deliberative discourse on educational innovation and change with accomplished and expert teachers, and other teachers, such as beginning teachers and student teachers. Each case study sought, within its own cultural context, to build capacity for educational innovation and change through opening a deliberative discourse between a variety of actors so that teachers could justify, communicate and re-conceptualise their pedagogical and professional practices.

A) Developmental case studies

In this section six developmental case studies with practical support, use of new materials or deliberative discourse for educational innovation in schools or teacher education from different countries are presented. They differ from the analytical case study in section B, where the situation of teacher education in two German states was analyzed within an innovative framework of border crossing.

Chapter 3:
Ireland: reflective and collaborative mentoring as a deliberative discourse with teachers, teacher educators and others

Geraldine Mooney Simmie and Sancha Power

Faculty of Education and Health Sciences, University of Limerick, Ireland

In 2006, at the start of the GIMMS project, there were two national reforms in Ireland seeking to influence science and mathematics teaching and teacher education. These interventions included syllabus revision, in-service education and training of teachers, grant assistance for improved infrastructure in schools and changes in science assessment in lower secondary education. One reform was concerned with interventions to improve the uptake of science and mathematics subjects among pupils (*Report of the Task Force on the Physical Sciences*, 2002). The second reform was concerned with the enactment of a legislative framework, the *Teaching Council Codes of Professional Conduct for Teachers* (2007) advocating for teachers to become publicly accountable within a self-regulating body with its own internal codes of professional conduct and standards.

At this time Ireland was experiencing a declining interest in the overall number of students choosing upper secondary physics and chemistry and theoretical mathematics (higher level mathematics). Gender issues were most noticeable at upper secondary education. Uptake of chemistry and mathematics was becoming co-equal in terms of gender uptake but overall uptake was well below national expectation. Senior cycle physics continued to be chosen by a majority of boys.

In this chapter we report on the background and context for science and mathematics teaching in Ireland during the lifetime of the GIMMS project, 2006–2009. We outline key features of the case study, data collection and analysis procedures, and the findings that emerged from the GIMMS Ireland. Finally we discuss some implications for policymakers, teacher educators, school leadership and teachers alike.

Background and context for GIMMS Ireland

At the start of the 2000s while there was some evidence of teachers committed to interrogating their practice and working in collaboration, the dominant discourse in schools appeared to mostly favour examination-success and teacher-centred teaching (Hogan, Brosnan & de Roiste, 2008; Mooney Simmie, 2007a, b). In general it appears that teachers were more familiar with lower-order cooperation, such as sharing textbooks, resources, strategies and information about pupils, in preference to higher order cooperation which included reflective approaches to thinking about their practices, peer-to-peer observation, mentoring or collaboration based on a dialogue of curriculum, innovation and inclusion (OECD *Teaching and Learning International Study, TALIS,* 2009a, b). Transmission models of teaching were commonplace with teachers working in isolation at the school site as the preferred option. Educational innovation and change in GIMMS Ireland involved teachers stepping outside the isolation of the classroom and deliberately engaging in a purposive and critical dialogue with colleagues, with teachers from other schools, with teacher educators and, at times with policymakers, on all aspects of curriculum development and pedagogy (OECD *Teachers Matter,* 2003).

PISA 2000–2003 findings on scientific and mathematical literacies showed a decline from average to below average standards, indicated the need for changes in the syllabus and the teaching and assessment of science and mathematics (Cosgrove, Shiel, Sofroniou, Zastrutzki, & Shortt, 2005). Some changes to syllabus and assessment approaches were adopted. In the early 2000s junior cycle syllabuses in science and mathematics were revised in lower secondary education (*DES Science Syllabus, Mathematics Syllabus*). For example, science became a more hands-on subject with pupils mandated to complete a number of laboratory experiments. Changes in mathematics focused on approaches to teaching for understanding. Assessment of junior cycle science, which traditionally involved a three hour externally evaluated written examination for all, 55,000 pupils approximately, was changed to a 65% written examination with additional marks allocated for practical work (10%) and investigation work (25%).

These syllabus changes were supported by in-service training of experienced teachers. This support was given in two phases. In the first phase teachers had access to a national support service which provided training workshops, school visits and on-line resources for a three year period, the *Junior Cycle Science Support Service* (JCSSS). The support service personnel, located within a regional education centre network, consisted of experienced teachers, seconded from their schools and working in cooperation with the inspectorate of the Department of Education and Skills (DES). This early implementation phase ended and was replaced by a much reduced generic support service. This generic service, the *Second Level Support Service* (SLSS), was responsible for supporting the implementation of reform measures in areas such as teaching, learning, discipline and assessment. In a similar way mathematics teachers had access to a similar support service, the *Junior Cycle Mathematics Support Service*. At this time a new pilot study *Project Maths* was developed to continue to assist mathematics teachers in a small number of schools (Ní Riordáin & Hannigan, 2009).

However introducing educational innovation and change onto the agenda of schools was far more complex a process than merely showing teachers newer methods of teaching and learning (Mooney Simmie, 2007a, b). New methods of teaching were less interesting to teachers who were coping with discipline issues and increased levels of bullying among young people. For example, the report of the *School Development Planning Initiative* (2002) showed that schools and teachers ranked support with discipline and bullying among their primary needs and far higher than any perceived need for dialogue on teaching, learning and reflection.

An additional constraint was the nature of the final written examination. Preliminary analysis of written examination papers for the science programme, by the authors of this chapter, showed a high emphasis on testing rote learning and memorisation, requiring single word key answers and leaving little or no space for depth of explanation or engagement with key principles of scientific thinking. The task of teacher educators and others to develop capacity with teachers for higher-order thinking was going to offer a real cultural challenge within this contextual landscape.

GIMMS Ireland

GIMMS involved teachers and teacher educators in a mentoring framework that crossed borders between school and university and was reflective, collaborative and inquiry-oriented. The social justice and inclusion focus in the project was on advocating for gender-awareness. This highlighted the need for science and mathematics teachers to be responsive to the varying needs of boys and girls in their classrooms.

We sought to generate a sense of ownership of the project at the school site and in a number of different settings. After an initial period of consultation we found schools and teachers who were willing to engage with their colleagues, and with the Department of Education and Professional Studies at the University of Limerick, to develop a collaborative, experimental and reflective approach. The work progressed in stages, including firstly developing the principles, practices and products that shaped *GIMMS Ireland* and finally evaluating and disseminating the findings to a wider audience of teachers, teacher educators and policymakers.

Framework for GIMMS Ireland

The framework for GIMMS had been agreed at the first meeting of the European partners (Chapter 1, Figure 1). Throughout the lifetime of the project GIMMS Ireland continued to develop an evolving theoretical framework. This was based on insights of key researchers and thinkers in the areas of educational innovation and change including Maynard & Furlong (1995), Brookfield (1995) and Darling Hammond & Bransford (2005).

Maynard & Furlong (1995) has already been dealt with in Chapter 1. Their critical co-inquiry model of mentoring was the preferred option for GIMMS Ireland as it was not based on a reproductive frame and held the potential for real engagement and mutual learning for all stakeholders. Brookfield's (1995) seminal text on becoming a critically reflective teacher became the preferred literature for the project team. He argued that one develops as a critically reflective teacher through a willingness to interrogate practice through multiple perspectives or lenses. These lenses he identified as the perspectives of pupils, perspectives of colleagues acting as trusted critical friends, the perspective of the research literature and the reflexive perspective gained from self-evaluation. Critical thinking in this way was framed as something that included both the wisdom of a reflective practice and the research literature. The key principle of engaging in this reflective, collaborative and inquiry-oriented way was found to be excellently framed within Darling-Hammond & Bransford's (2005) professional practice model. This model considered professional practice as requiring knowledge of subject matter, knowledge of the learner and learning and knowledge of assessment.

This evolving theoretical framework, for GIMMS Ireland, assisted in the development of definitions of innovation, mentoring and gender awareness and gave greater clarity to the project goals:

- Educational innovative practices were interpreted as eliciting higher-order cooperation among science/mathematics teachers for the development of pupil-centred curricular practices. For example, teachers worked together in GIMMS Ireland in public incubation spaces that focused on a deliberative discourse on curriculum, learning and innovation within their school, across different schools and with the university.
- Mentoring was interpreted as a reflective relationship of learning between teachers, across their professional lifespan, and across borders between teachers and teacher educators with interchangeable novice-expert roles for the purpose of mutually enriching co-learning.
- Gender-awareness required that teachers and teacher educators consider gender as a specific criterion, when co-planning, designing materials, teaching and evaluating teaching to ensure that both boys' and girls' learning needs were included in the science and/or mathematics classroom.

Data collection and analysis

The research approach used a case study paradigm. Data collection methods involved triangulation of perspectives including the perspectives of teachers, pupils and teacher educators/researchers (Merriam, 1986). It was exploratory in design and capable of offering fresh insights into the possibilities and constraints of developing newer and innovative approaches for teaching for pupil learning in the dynamics of the public incubation space that was created and designed to generate educational innovation and change. Data sources included surveys, reflective journals by teachers and their pupils, focus group interviews and field notes taken at key stages during the lifetime of the project.

The main research and development tools for the project, reflective diaries, surveys, focus group interviews and evaluation instruments, were developed in partnership between the schools and university. Data analysis involved a constant comparison approach to the selection of themes and independent validation, between the two researchers, while using the theoretical framing for the project.

GIMMS findings

Findings from each of the phases are now considered giving a range of practices, perspectives and products from initial contact with teachers, to reflective journals of pupils to the voice of the teacher as captured in the focus group interviews.

Initial contact with schools and teachers

Seven schools were initially invited to partake in the GIMMS project, two of which withdrew at the beginning of the project. One school later withdrew while the project was in progress. The reasons behind withdrawal from the project were varied but all appeared to be concerned with the business of the school day and the lack of clarity from the project team at the university. One school felt that timing was an issue and a reason for not becoming involved. One science teacher felt that he/she

"had enough on her plate trying to find everything, and was not hard pressed to engage in a research project with the University of Limerick".

There seemed to be a lack of motivation regarding working collaboratively and becoming part of the project in general, a sense of "my work is good enough already, it doesn't need to be improved" was felt and noted (Field Notes, 17/10/2007). Along with this lack of enthusiasm there was a sense of the business of the teacher workload and the lack of a reflective stance:

"Look around you there, see all that equipment, we are supposed to get all that ready and tidied away before the next class, do you know how long that takes". (Field Notes 28/02/08 Teacher comment)

One of the schools withdrew some weeks into the project. They agreed to take part in a debriefing seminar to give insight into their reasons. It became apparent that there had been a lack of communication within the school and with the university. Some staff knew little or nothing about the project and felt that they had missed an opportunity: "that is very interesting, it's a shame, I would have loved to identify how my students learn and work best" (Field Notes 28/02/08). The project team learned the value of not making assumptions, checking back with teachers and schools on all protocols and procedures and providing greater clarity in the early phase about the type of commitment level required.

GIMMS schools and teachers

After the initial phase of consultation four schools eventually became involved with GIMMS and remained with the initiative to develop innovative practices. Three schools developed new approaches and practices in their science teaching and one school developed innovation in mathematics. All schools involved came from the mid-west region of Ireland, which comprises Limerick, Clare and North Tipperary. Each school involved was deemed to be a 'large' post primary education provider. They were all co-educational. This is quite important as just 477 of the 732 post primary schools listed within the Department of Education Statistics for 2006/2007, were categorised as co-educational (Central Statistics Office, 2007). Ireland has a strong tradition of single sex education. Most schools worked with either 1st or 2nd year pupils. Table 1 gives background information on the GIMMS schools and their area of investigation. School types were represented with one community school, one fee paying voluntary secondary school and two community colleges.

Table 1:   Profile of GIMMS Ireland schools

Name of School	School Background	Lab Technician	Student Numbers	Mixed Ability	Topic for GIMMS
Oak	Suburbs	Yes	650+	Yes	Micro Scale Science
Ash	Suburbs	No	700+	Yes	Data logging in Science
Elm	Urban/Rural Mix	No	700	Yes	Promotion of ICT Enhanced Learning
Elder	Urban/Rural Mix	n/a	650	Yes	Incorporation of ICT into Mathematics Teaching

The GIMMS teachers, in consultation with the teacher educator, opted for different topics in science and mathematics. They planned to explore these in a collaborative and reflective way. OAK explored micro scale science using the stimulus of some mandated pupil experiments in the new syllabus. Their rationale for this type of laboratory practice included three main perspectives: for the improvement of pupil motivation, for health and safety considerations and for a 'greener' approach to using and disposing of waste chemicals and materials. Science teachers in ASH had previously attended national in-service education and training in the use of data logging equipment. While the in-service had provided some support to them they still felt that they were not fully confident in using this equipment. They wanted GIMMS to assist them in making better use of the equipment in their teaching of science for understanding. Science teachers at ELM had, in their new state-of-the-art laboratories, access to new ICT technology to improve presentation of lesson materials. They were not using this technology to full advantage. They wanted through GIMMS to explore differences in pupil learning between traditional "chalk and talk" approaches and ICT-enhanced approaches to teaching instruction. Mathematics teachers in ELDER also wanted to explore teaching for understanding and the use of ICT-enhanced learning activities in the mathematics classroom.

GIMMS pupils' perspectives

Pupils in the case study schools took part in a survey before and after the project. The demographic of pupils involved in the survey was predominately female, with an average age of 13 years. All pupils belonged to the first year cohort in each school and the majority had completed science to some degree at primary level. In terms of a future in science the study cohort was quite consistently positioned in a designated "don't know" category prior to project. Within the project pupils took part in two aspects of data collection – a pre survey and post survey of "attitudes to science" and the keeping of a reflective diary.

The attitudes of pupils towards science were somewhat similar across the four schools pre and post implementation of the project. Most pupils agreed to like science while acknowledging a strong dislike for mathematics. This dislike for mathematics was further reinforced when "maths is all theory and no fun" was very strongly agreed with.

In terms of the preferred teaching and learning methods, again there were strong correlations between the four schools. There were strong preferences for the integration of ICT, coloured posters, videos, DVDs and diagrams to improve learning. All pupils surveyed agreed that working in groups aided their learning. Pupils were in favour of linking new material with old material and also liked when new learning was related to everyday life. Pupils agreed that mathematics was easier when explained through real life examples and learning was enhanced when done through activities. There was a strong preference from pupils also towards being actively engaged. In the section on 'my learning' pupils acknowledged that they learn more by doing, and that carrying out investigations was fun. In relation to predicting results before doing activities and experiments, pupils "didn't know" if this was helpful. Perhaps they had no experience of using this approach. Pupils also identified that they liked receiving feedback from the teacher on a one to one basis.

Attitudes to teaching and learning of science and mathematics did not change in the post test surveys completed by the pupils. To investigate further the impact of the curricular tool, the post surveys contained additional questions unique to each school's project. This was where pupils' perspectives were most keenly seen in relation to the project and developments made by the project team of teacher and university researchers. There was consensus across the four schools from pupils that they enjoyed the interventions that had taken place within the classes over the past few weeks. Common themes which emerged were concerned with how interesting they found the new approaches and how much easier it was to understand the subject matter presented. Comments across the schools in relation to why they liked the new approach inspired answers such as "interesting" (all), "fun" (all), "colourful" (ELDER), "easier to understand" (ELDER), "liked using modern equipment" (ASH) "because it helped me and the rest of the class learn better"(ELM).

Those that used data loggers acknowledged that it was "easier to get a graph", "helped to learn", "something different". "Fun, didn't realise I was learning". When asked to identify key words which described the activities done, students circled "cool", "enjoyable", "fun" "difficult". They were quick to acknowledge that there was some level of difficulty in using the new equipment. But overall, datalogging was viewed to be a success in integrating pupils with both science and mathematics data, "does engage them more in that they are I suppose because they are using this technology it looks more like a Nintendo DS and it gets them in that way" (ASH).

ELM had carried out a project also related to ICT and reported very positive results. The science teachers planned to continue to use transmission models of teaching but to use more visually attractive ICT ways of presenting the subject matter. The cohort of pupils involved identified a strong preference for the more enhanced ICT classes. When asked to comment why, statements like "it's easier to understand", "it is better than taking notes" "its much more interesting" "better than the book" were found. However, there were a few that acknowledged dislike for these new presentation methods; they preferred actively doing their own experiments "no I like experiments and doing things myself".

Those involved in the micro scale project indicated mixed feedback, some students really enjoyed the new equipment "interesting", "cool", "enjoyed it", "cute", "less mess" while others remarked how they found it "difficult working with the small apparatus", "made your work harder, difficult to pour in liquids", "easier to break". This was strongly endorsed by the teachers when we met to share and reflect with colleagues, "the (pupils) are tuned into it because it is kind of like a television programme, it's CSI," … "They love it, its ultra modern, it's not clunky" "however, some especially the boys find that their hands are big and awkward … you have this big (boy) who is 14 but nearly six foot and his hand is this size and he twists it and the whole neck comes off the bottle (OAK).

Possibly the greatest level of change of perspective was reached within the mathematics project in ELDER. The new approach generated huge interest among the pupils. Pupils remarked how there was "nothing to dislike" about the new approach. They found the new approach more "interesting", liked it because it was "more colourful", "she explained it more", "because it was new", "a new way of learning". The mathematics teacher remarked on the high levels of motivation and new found enthusiasm and even a remarked change in classroom behaviour which she was hoping to achieve through the project (ELDER).

While innovative teaching and learning approaches were found to be a positive experience by pupils they were less complementary of having to write a reflective diary. There was a strong negative disposition toward diary writing from all pupils. Most of the pupils saw them as *"no point"*, *"annoying"* and *"boring"*. However one or two pupils in each school acknowledged a positive outcome in the way in which the reflective diary *"made you think, "let you express your feelings", "helps the teacher"* and *"because you know how far you have gone"*. This experience of diary writing as part of a science or mathematics classroom was a new experience for pupils and for their teachers. It challenged both the pupil and the teacher in terms of thinking as well as improving their capacity for writing and literacy.

Teachers' perspectives

The teacher autobiographies drew on three aspects of data collection – teacher reflective diaries and focus group interviews both of which were further supported by field notes taken by the researchers. On analysing the findings some common themes emerged which may be considered in terms of a dynamic force field. What were the forces propelling and motivating teachers forward towards innovative and inclusive practices and what were the resisting forces pulling the teachers back to more traditionally inherited approaches to teaching, learning and mentoring?

Most teachers in the study made reference to time and workload as strong resisting forces with one teacher noting that this approach was not supported by the dominant culture:

> "It took us too much time to cover the material, it took a couple of class periods that we didn't really have for students to complete their questionnaires. It took us as teachers quite a lot of time to prepare material and assess it and all that sort of stuff so the time factor was definitely a negative and I would be including the reflection in that". (ELM)

> "It is written somewhere, somewhere in official documents it is written in small print that new methods are to be recommended but (while) it is written in small print, you don't hear it spoken, it's not really supported". (OAK)

GIMMS Ireland teachers engaged with the collaborative planning and reflective approaches with colleagues, pupils, teachers in other schools and with teacher educators and researchers. All teachers reported high levels of motivation while carrying out the project. This was strongly supported within focus group interviews where teachers commented that they had gained significantly through collaborative reflective dialogue about teaching for learning and learning new skills sets with colleagues and using ICT equipment in the school, *there was a lot of skill learning for teachers* (ELM). While teachers appreciated working together at school they appreciated the broader lens of working with other schools and the university:

> "It is very good to see outside our own little 'world' and to feel that we are engaged in something bigger than what goes on in our own classrooms because teaching in its very nature can be quite an isolating job". (ELM)

Some teachers were motivated by the way both themselves and their pupils were being "pushed" into learning "how to learn":

> "In my view one of the important things for students to learn as they go through secondary school is how they learn, and I felt that in engaging in the questionnaires and their own reflections that it began to develop how they did their own learning". (ELM)

> "This project forced us into doing it in a more analytical way, the whole process of reflecting and particularly as regards our methodologies". (ELM)

Teachers were also motivated by such practical considerations as developing newer resources for immediate classroom use and one teacher stated that "sharing confirms best practice":

> "We are devising some worksheets and these worksheets are going to be quite useful every time we use the lab quests so a few of us are devising worksheets and these are going to be available for the rest of the department so it is sharing resources as well". (ASH)

The GIMMS approach helped *sustain (pupils) concentration in class* (ELDER) while another teacher noted the benefit of being challenged in their thinking:

> "Things like collegiality and being a reflective practitioner because we don't get time to write down notes to mull over what goes on in every class so this is definitely an opportunity to put a lot of thought into what we are doing and how we are doing it". (ASH)

The critical factor that seemed to constrain was the emphasis that all teachers gave to the busy environment that was 'teaching to the test':

> "(I'm) concerned as to how this would relate to exams and application to exam questions in terms of marking schemes and how they will compare and we would be a little bit worried about examiners … maybe if a student wasn't

exactly clear on what they did would they even be able to get the message across about how they completed the experiment on a mini scale to a recognisable level on the exam". (OAK)

In GIMMS Ireland there was largely gender silence from science and mathematics teachers. Teachers of these subjects appeared willing to engage with this issue at a rather superficial level, for example, using direct examples of boys and girls in visual aids. However teachers were reluctant to engage with this topic as a deeper level of discourse on this topic. It may be because science and mathematics teachers in Ireland have had a strong tradition of perceiving themselves as purveyors of subject matter knowledge and share less of a perception of their activist professional role as agents of change for social and political change (Sachs, 2003). Additional sustained supports will be required if science and mathematics teachers are to become willingly to critically engage with this political and ethical aspect of teaching.

Products, practices and dissemination

GIMMS Ireland developed a number of new products and practices at the case study schools. These included a *Resource Booklet* of classroom resources and collaboratively designed *Reflective Diaries* for both teachers and pupils. A *Data Logging Resource* manual was written and developed in one of the case study schools. All products were trialled and tested with other experienced science and mathematics teachers and student teachers.

Evaluations showed that the resource materials generated were valuable learning aids for pupils and increased the motivation of both pupils and teachers. They included visual learning aids, aids for micro scale practical pupil experiments and ICT-enhanced instruction and presentation. They also included reflective journals written by both teachers and pupils. While teachers and pupils engaged with the journals there was evidence that this process strongly challenged both teachers and pupils. The reflective process and reflective writing in particular, is known to have the capacity to develop deeper levels of higher order thinking skills and improve levels of literacy and depth of analysis (Bolton, 2010). This aspect of GIMMS Ireland was a new demand for experienced and accomplished science and mathematics teachers.

GIMMS Ireland results need to be viewed within the larger national background and context. The project succeeded in motivating science and mathematics teachers, and their pupils, to engage in writing reflective journals and sharing reflections about teaching, learning and innovation with each other and with teacher educators, and sometimes with policymakers. It was less successful in generating deeper levels of critical reflection and analysis. The depth of the cultural task involved in challenging and supporting experienced teachers move away from traditional teaching approaches is underestimated in national reform measures.

The reflective mentoring between the school-university partnership and within the classroom, between teachers and their pupils, was judged to be a success as a starting position for a deliberative discourse between teachers and teacher educators. The depth of collaboration and reflection engaged in during the various settings and groupings within the project was a new and innovative experience for all teachers in the project. It took place against the backdrop of largely transmission teaching, resistance to teacher writing and an examination system that, at the time, mostly favoured lower order questioning. The findings from GIMMS Ireland, including the pupils' and teachers' perspectives, show that public incubation spaces, like the ones provided in the project, can bring real change to a teaching team within a school through offering consistent levels of both internal and external support and challenge. This juxtaposition of internal and external discourse developed educational innovation and change within the project as critical thinking that supported experimentation and the development of creative ideas and practices.

GIMMS Ireland products, protocols and practices were disseminated to teachers, experienced teachers, beginning teachers and student teachers through a number of pathways. They were disseminated through the GIMMS web-site http://www.gimms.eu (GIMMS 2006–2009) and through the publication of the *Physical Science Journal* (2009) which was issued in electronic format to all science teachers. They were disseminated to teachers in the mid-west region of Ireland through two national conferences, on the 14th January 2007 and 25th/26th September 2009 and through a poster presentation at the *Chem-Ed Conference* (ChemEd-Ireland, 2009) and the *Department of Education & Professional Studies* in the University of Limerick in 2009. GIMMS Ireland resource materials were disseminated to over two hundred second year student teachers in 2008/2009 at the University of Limerick. They were disseminated to teachers in twelve other European countries through networks and exchange made possible by connection with GIMMS transnational partners and another Comenius 2.1 project, CROSSNET Ireland (Mooney Simmie & Power, 2012).

Sustainability

GIMMS Ireland helped teachers develop the capacity to engage in a deliberative discourse at their school, with other schools and with a higher education institute through a school university partnership dedicated to educational innovation and change. GIMMS opted for the development of a reflective and collaborative framing for mentoring as the preferred way to engage teachers and teacher educators, and sometimes policymakers, as partners in educational innovation and change. The project took place, at the same time, as the development of an academic pathway for a *Masters in Educational Mentoring* programme at the University of Limerick (Mooney Simmie & Moles, 2011).

Discussion and implications

GIMMS Ireland achieved a number of successful outcomes and identified some constraints that move science and mathematics teachers beyond traditionally inherited practices. It developed a reflective and collaborative framing for mentoring that generated a deliberative discursive inquiry on pupil learning, innovation and inclusion. This deliberative discourse crossed borders between a number of different communities, teachers, teacher educators and others. This extended network supported teachers' critical engagement with their science and mathematics teaching through professional networks at the school site, beyond the school and with other schools through school-university partnership. As argued by Brookfield (1995) it is this multiplicity of border crossing perspectives and actors that is at the heart of becoming a critically reflective teacher. It was this variety of actors, and especially teachers interacting with other teachers, such as student teachers and beginning teachers, and teacher educators that gave experienced teachers access to research findings and new, innovative and ICT-enhanced approaches to teaching and assessment.

Innovative products and practices included the generation of critical thinking and creative thinking through public *incubation spaces* for teachers to exchange with new ideas and practices with each other as a school team, with colleagues from other schools and with teacher educators/researchers from the university. These *incubation spaces* consistently supported teachers' ideas and activities as they tried to grapple with new ways of teaching that went beyond traditional pedagogies. Educational innovation and change in the project was realised through these multiple settings and was most noticeable in the reflective writing engaged in by teachers and their pupils, in the creation of new laboratory systems for pupil experiments, and in the development of hands-on activities and ICT-enhanced activities for classroom use. It was also evident in the planning developed within the school-university partnership and the level of higher-order cooperation between teachers in the project (OECD TALIS, 2009).

Pupils' perspectives in GIMMS Ireland showed evidence that new ways of teaching, and more engaged teachers, were found to make science and mathematics motivating and enjoyable while leading to improved understanding. This development was also observed in the school even when ICT was used largely to support teacher-centred models of teaching. Pupils' perspectives were positive in both cases.

The pupils in GIMMS Ireland, and their teachers, were less supportive of having to engage with reflective diary writing and with learning about science and mathematics. Reflection is known to be a valuable tool for learning 'how to learn' and to improve thinking capacities (Bolton, 2010). Reflective practice and critical thinking clearly needs further support and persistence if deeper levels of innovation are to be achieved in the classroom and school. In this project it was a new experience for both teachers and pupils. If further developed this border-crossing deliberative discourse framing for educational innovation and change could become a valuable counter-weight to an overly busy school culture and assist the development of capacity for critical thinking and literacy among both teachers and pupils.

Teaching in innovative ways challenges the mind-set, values and existing traditional culture and dominant discourse of schools in Ireland. In GIMMS Ireland teachers remained highly motivated throughout the project and engaged fully with a range of actors including colleagues, teachers in other schools, the university personnel and others, including a number of national policymakers (Mooney Simmie & Power, 2012). They reported high levels of skill and competence learning, especially in the area of ICT, through this collaborative, border crossing reflective platform. GIMMS Ireland had success in developing higher-order cooperation among teachers, removing teacher isolation and supporting collaborative teacher-teacher interaction and dialogue on curriculum matters.

Teaching to the test proved a consistent constraint to the development of educational innovation and change. GIMMS teachers constantly reported their concern that they might be failing their pupils by focussing on generating innovation in their classrooms rather than giving sufficient time to delivering 'right answers' to state examination questions. This dominant discourse of right answers and success in examinations overshadowed the entire project at all times. These findings are particularly timely as Ireland reviews its junior cycle assessment system and seeks to improve literacy and numeracy among pupils and teachers (National Council for Curriculum and Assessment *Innovation and Identity: Ideas for a new Junior Cycle,* 2011a, b, c, and *Towards a Framework for Junior Cycle,* 2011, Department of Education and Skills *Literacy and Numeracy for Learning and Life,* 2011–2020).

GIMMS Ireland gave science and mathematics teachers in each of the four schools access to a broad-based democratic network of colleagues, teachers in other schools, teacher educators and national policymakers. The school principal, in each of the GIMMS schools, showed a deep interest and commitment in pedagogical educational innovation and change.

Initially GIMMS Ireland schools and teachers required detailed guidance on the aims and outcomes of the project. They also required considerable scope for curriculum making and innovation. The balancing of these two opposites, detailed guidance and curricular freedom, was necessary in giving GIMMS teachers the ownership and motivation to complete the project successfully. It is clearly no longer sufficient to leave teachers, on their own or within their own professional networks to become curriculum innovators as envisaged by Stenhouse (1975) and others, during the golden age of curriculum making in the 1970s. Neither is it sufficient to see this as a top-down training problem of skill, competence and disposition. GIMMS argues that educational innovation and change requires a new approach from all education actors – a boundary crossing collaborative discourse between teachers, teacher educators and, others such as policymakers.

In conclusion, as long as teaching to largely lower-order tests is perceived as the route to pupil progression, and continues to be supported by policymakers and society generally, then trying to bring about educational innovation and change in the classroom and school will continue to remain a demanding, difficult and counter-culture movement. It may be delusional to think that we can continue to do both simultaneously:

> "(We have a) shared understanding of education as an enlightening and emancipating force for the democratic development of each person. We have remained acutely conscious of the struggle to retain this conception of education as a human liberating force against the backdrop of a reductionist agenda sweeping the education world with its focus on outcomes and external modes of accountability. In our opinion, mentoring in teacher education needs to be underpinned by an alternative lens provided by the literature taking this broader educational landscape into account." (Mooney Simmie & Moles, 2011, p. 471)

Chapter 4:
Denmark: empowering innovation through experienced teacher – student teacher democratic collaboration

Lotte Skinnebach and Birgitte Stougaard
University College Lillebaelt, Department of Teacher Education in Jelling, Denmark.

> "For the first time in my long career as a mentor for student teachers, I was challenged to go deeper into technical and pedagogical issues. Until now our discussions had primarily been of practical issues".
> (Experienced school-based mentor teacher commenting on GIMMS Denmark)

Background and Context

During the first decade of 2000 the findings of the PISA investigations were taken very seriously by decision makers in Denmark as they were in many other countries. While Danish students in mathematics achieved slightly better results than the OECD average, the results for science, especially in PISA 2000 and 2003, troubled political leaders and policymakers.

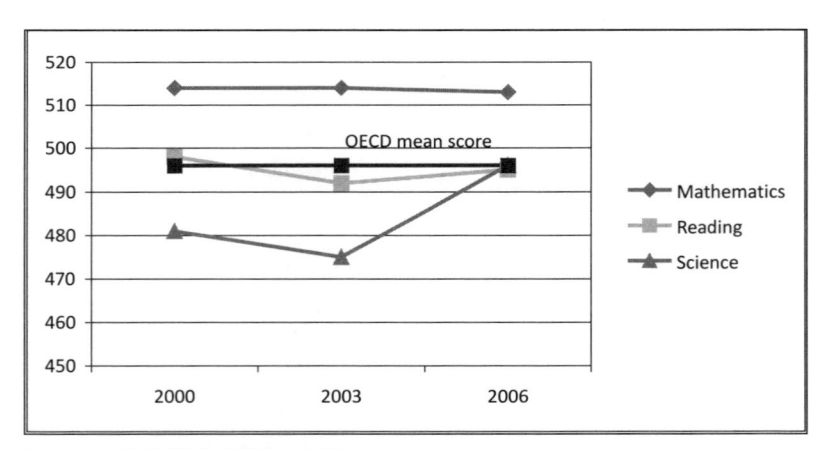

Figure 1: PISA 2000, 2003 and 2006

Danish mathematics students showed constant achievements in the PISA investigations (Figure 1, Egelund 2007; OECD 2001, 2004, 2007). In both PISA 2000 and 2003 the result was a score of 514 compared with the OECD average of 496. In PISA 2006 the mathematics students achieved 513 points. In PISA 2000 and 2003 science students performed below the OECD average of 496 points with 481 and 475 points respectively. In PISA 2006 the result for science was marginally better with 496 points.

Seen in a Nordic perspective, there was a significant difference between the achievement of boys and girls in both mathematics and science. The difference was 9–10 points in both subjects with boys achieving better than girls. In Finland and Sweden, as in most of the OECD countries, the gender differences in achievement appear to be insignificant. In Iceland and Norway the small difference is in the girls' favour, with Denmark as the only Nordic country where boys continue to outperform girls in mathematics and science (Sørensen & Andersen, 2007).

Compared with all OECD countries, Danish students have the least positive personal attitude towards science (Sørensen & Andersen, 2007, p. 105). They do not see science and technology as a way to social promotion, but they are positive towards the importance of science in society. While students in the Nordic countries in general value science less than the OECD average, the Danish students value science the least. While geology has the least interest followed by chemistry and physics, the Danish students tend to be more interested in human biology. The findings in the ROSE investigation underline the differences regarding gender and interest in science (Sørensen, 2008). Girls and boys appear to be interested in different aspects of science, and while girls generally seem more interested in human biology and health, boys appear to be more interested in technology and physics, for example, nuclear weapons.

PISA results influenced policy decisions in Denmark in a variety of ways. Mathematics and the sciences were given more lessons: one more lesson per year in mathematics in 1st to 3rd grade (6-8 year-olds) in 2003–05, physics/chemistry had an extra lesson in 9th grade (2005), biology was extended with a year in the 9th grade (2005), and so was geography one year later (2006). From 2009 science as an integrated subject was introduced in the optional 10th grade.

In 2007 mandatory external examinations were introduced in a wider range of subjects (for example, biology, geography, physics/chemistry) along with the three traditional examination subjects – Mathematics, Danish and English. Subjects for the final examinations (Skolestyrelsen, 2009) are decided by a lottery system and some examinations/tests (biology and geography etc.) are taken on-line. New obligatory national tests were planned at different age levels in 2007–2009 but were postponed. The tests included e.g. mathematics (3rd and 6th grade), biology, geography and physics/chemistry (8th grade) (Undervisningsministeriet, 2009a).

In a three-year period, from 2007–2010, schools received in-service course-fee compensation when enrolling teachers on in-service courses in mathematics and science subjects, including nature/technology, biology and geography. The in-service courses were at different levels, for example, subject specialization and diploma level.

In addition *National Centres of Education* have been established: NAVIMAT[1] for mathematics in 2007–2011, and two initiatives in science, CAND in 2006–2009 and NTS[2] in 2009. A new and more detailed curriculum for all subjects in the Danish comprehensive school was introduced in 2009 (Undervisningsministeriet, 2009b). A new act on teacher education was introduced in 2007. One of the foremost ideas behind the act was to strengthen the choice for the sciences by allowing specialization in two to three science subjects. The new act also emphasized the importance of specialization at different age levels, for example, mathematics education for primary education and for lower secondary education respectively. As a consequence of this new act there have been problems recruiting student teachers with a specialization in science.

Another factor worthy to note is the qualification requirement of teachers. In Denmark teachers for upper secondary schools are educated at universities, and qualify at Masters' level. The teacher education and qualification requirement for primary and lower secondary schools is a four-year Bachelor programme which until now has taken place at university colleges, such as University College Lillebaelt.

The GIMMS model

The model in Figure 2 was developed at a meeting for the GIMMS coordinators as a model to explain the role of actors involved in the project and the rationale and thinking behind the project. It was essential for the GIMMS team to make a model that could work for all countries – this gave a coherent framework for the project and allowed the 'space' for each country to interpret the details to suit their own context and to respect the 'diversity' of voices in this regard. These underlying democratic principles set the background and context for the GIMMS interpretation of innovation principles, practices and products.

The model was a crucial touchstone for the Danish track. It was used throughout the project to explain the different mentoring relationships of learning within the project and the collective 'view' of the GIMMS team on educational innovation and change. While we used the model in Denmark we also extended it to suit our cultural context.

We wanted to emphasize that teaching approaches for pupil learning needed to focus on gender-awareness and students' interest and motivation. This was our interpretation of innovative practice and the way to bring this innovation about was through these mentoring relationships of teacher education, development and inclusion. Research studies show that teaching must inspire students through a diversity

---

1   http://www.navimat.dk/
2   http://www.nts-centeret.dk/

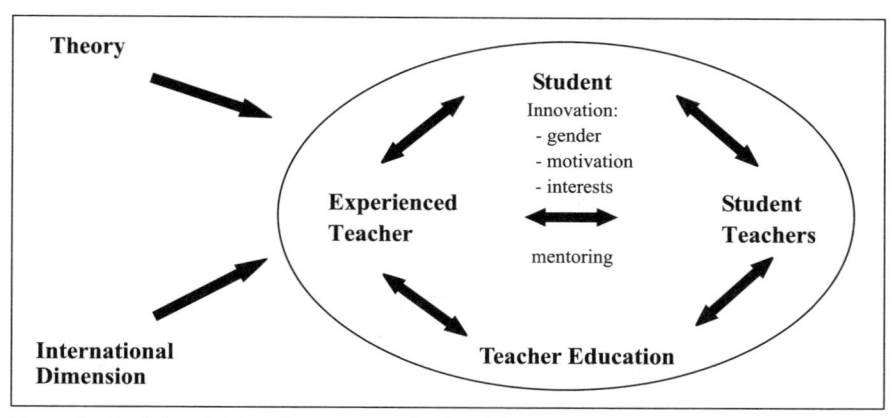

Figure 2: Model explaining the actors involved and the rationale of the project

of perspectives on themes, context based education, variation of tasks and activities and evaluation methods (Heuvel-Panhuizen, 2006).

The GIMMS model shows that through an extended social democratic network, bringing teachers from their private to a more public world, through a variety of actors and with access to research findings, that theory can contribute to developing a deeper understanding of the subject matter knowledge and knowledge of how to teach. One of the aims of GIMMS was to connect theory and practice through this extended democratic mentoring platform. Furthermore, we wanted the model to illustrate that throughout the lifetime of the project, 2006–2009, the European collaboration with GIMMS colleagues across five other countries influenced this agreed educational innovation and change perspective and practice.

The Danish perspective on innovation

According to the Danish GIMMS model, educational innovation and change was defined as developing mathematics and science teaching and teacher education by focusing on pupils' motivation and interests in the subjects on the one hand, and gender awareness on the other hand. Gender-awareness was interpreted to mean equal opportunities for boys and girls in science and mathematics classrooms.

In Denmark, even though teacher educators and researchers within mathematics have tried to change mathematics teaching for many years, there are still a considerable number of teachers who teach in a very traditional way. Teaching is mainly teacher-centered focusing on solving mathematics tasks. The teacher shows an example of how to solve the task, and students replicate this example using a formulated approach. Of course there are teachers who choose different approaches, but the former is known to be the primary pattern for mathematics lessons in Denmark.

Regarding science education, the importance of experiments is widely accepted in Denmark. It seems that there are difficulties in connecting the findings of experiments with theory and with full scientific understanding of key concepts and principles. Whole-class instruction is still very common in the science classroom. However, there is a growing interest in the use of active collaborative learning, ICT-enhanced teaching and learning and using different kinds of pupil presentations as pedagogical tools to access higher-order thinking. Research shows that this leads to better scaffolding and connection between the practice of 'hands-on' experiments and the theory, namely the scientific knowledge and principles.

This contextual background gave the point of departure for discussing GIMMS innovative practices in science and mathematics teaching with experienced teachers and student teachers. As a result of the discussions, the experienced teachers and student teachers defined innovative practices as: "different ways of organization rather than whole-class instruction – other ways of organization than one teacher addressing a whole group of students". It also meant new teaching materials and an inductive approach to teaching – from practical experiments and activity based learning to general theory – in addition to a deductive approach. It was also decided that different ways of organizing the learning environment needed to include gender awareness.

With respect to the Danish tradition of democratic collaboration, and the development of ownership of the project by the GIMMS teachers, the project coordinators accepted the definition above and worked alongside teachers to realize these aims and outcomes. However we expressed our concern that gender-awareness should be included, not only in ways of organization of the classroom learning environment, but also according to teaching materials and the choice of different teaching topics.

The democratic mentoring platform in GIMMS Denmark

The GIMMS model focused on a number of different mentoring relationships in the Danish part of the project:
- Student teachers and experienced teachers
- Experienced teachers and teacher educators and
- Student teachers and teacher educators

The importance of reflection within mentoring relationships had already been emphasized by Maynard and Furlong (1995). The mentoring relationships in GIMMS Denmark – based on reflective and constructive collaboration – were initiated in a workshop in the beginning of the project and continued through a web-based discussion platform (First Class Skolekom). The latter was established for ongoing communication and exchange of reflections.

The mentoring relationship between student teachers and experienced teachers was based on the idea of action research (Plauborg, Vinther, Andersen & Bayer, 2007) and the planning of the student teachers' practice period was the central part of this relationship, founded on mutual respect and equality rather than as a novice-expert mentor-mentee relationship.

The experienced teachers and the student teachers included their common plans in a contract for their collaborative work in order to clarify communication and to make their work mutually advantageous, thus highlighting the responsibility of both parties. To ensure that both partners agreed on the plans, the agreed contracts were returned to the GIMMS teacher educators and national co-ordinators.

In Denmark in-service courses on mentoring are available, but there are no strong traditions of any specific mentoring model. Experienced teachers remain private in their classrooms and do not like to be observed. While there is no formal hierarchy in the student teacher – teacher relationship, experienced teachers tend to observe from the back of the classroom, and are also responsible for the final evaluation and grading of the student teacher.

In the mentoring relationship between experienced teachers and teacher educators the focus was on educational innovation and change, according to the Danish GIMMS model. The collaboratively agreed agenda was on developing innovation of teaching and teacher education focusing on gender-sensitivity (Sørensen, 2008), student involvement and collaborative learning. The project challenged traditional teaching and the privacy of the teacher. The project was implemented by an extended democratic network that crossed borders and had mutual learning at the centre.

Gender awareness

In Danish the word 'køn' has a double meaning: it can be translated both into the biological term 'sex' *and* the sociological term 'gender'. The English term 'gender' is much wider and can be used to discuss cultural, social and economic differences. In Denmark gender awareness is often seen as a left-over from the women's liberation movement of the 1970s. The debate frequently tends to reduce the questions to whether education favours girls or boys and it becomes difficult to move beyond this reductionist approach to the issue.

However, it is not acceptable to reduce gender awareness to either a predetermined biological difference or simply a left-over from the women's liberation movement. The findings from PISA 2000, 2003 and 2006 show the differences to be too significant between the achievement of boys and girls as regards both mathematics and science education. In addition, the ROSE investigation shows that boys and girls seem to be interested in different themes of science and clearly gender is an issue of significance in science and mathematics teaching and teacher education and in cultivating teacher-awareness of pupil learning for all pupils, irrespective of gender (Sørensen, 2008).

In the Danish part of the GIMMS project, questions about gender awareness and sensitivity were introduced in a workshop by a specialist scholar from the University of Copenhagen, Inge Henningsen. Gender awareness was seen in a PISA context (Henningsen, 2005). Discussions among participating teachers and student teachers underlined the general idea in Denmark that gender is "not a big issue – we have already been there". The real issue seen from their perspective was pupil motivation and (lack of) interest. Somehow they did not associate gender with motivation and interest. When teachers and student teachers addressed gender issues in the project they tended to conceive of it as merely requiring them to divide the class into single sex groupings. This was a best fit approach to their concept of innovation as an experimental approach to changing the organization of the learning environment. According to the Danish GIMMS model and our understanding of innovative practice we worked with the project groups to extend this understanding of gender-awareness.

Cooperation and settings in the project

School – university college partnerships were established between four case study schools (Bakkeskolen, Skjoldborvejens Skole, Vinding Skole and Bredagerskolen) and the Department of Teacher Education, University College Lillebaelt. Heads of the schools signed a contract which set the conditions for future cooperation: The GIMMS project and University College Lillebaelt were responsible for workshops, seminars and mentoring of experienced teachers. The schools were responsible for mentoring of student teachers and cooperation with teacher educators. The schools offered teachers involved in the project twenty five additional working hours per year.

There were, however, differences in the involvement of the case study schools. Some of the experienced teachers played an active part in the student teachers' school practice. The mentoring relationship established was mainly between student teachers and experienced teachers. At all case study schools a mentoring relationship between an experienced teacher and teacher educators was established. In the first case study school the focus of the mentoring relationship (experienced teacher – student teacher) was on development of mentoring practices and innovative practices. As regards mentoring between experienced teachers and teacher educators, the focus was on the development of innovative practices that challenged traditional curricular approaches, for example, collaborative learning, designing effective work groups, selecting materials for teaching, gender issues and the use of ICT in pupil presentation.

Danish teacher educators are not specialist academic researchers. The main part of teacher educators' workload is assigned to teaching, and only a minor part of this can be dedicated to innovative research and developmental projects. However, one of the most important settings is that lecturers at Danish University Colleges in their research and/or developmental projects focus on Mode-2 science, where knowledge is generated in the context of application (Nowotny, Scott & Gibbsons, 2001).

Schools in Denmark are comprehensive schools from primary to lower secondary, encompassing the age range from 6 to 16 years. Teachers are educated to teach all groups of students and at all levels in their subjects. Because of that, there has been a long tradition of teaching in Denmark that is child-centered in preference to subject-centered. As mentioned in the beginning of this chapter, teachers in Denmark are facing new challenges. They now have to be more subject-centered while still bound by the tradition of child-centered teaching. The GIMMS project offered an opportunity for experienced teachers to reflect on the possibility of introducing new and innovative practices as well as discussing important changes in their professional life as described above.

Teaching science: the product perspective

The products produced in GIMMS Denmark can be divided into two groups where one is easier to describe by means of the process, and the other was disseminated as an example of gender-sensitive products for teaching science. This latter resource was inquiry-based material for 9th grade students in physics and chemistry.

With the aim of furthering motivation, student learning and the use of ICT, three different story lines for teaching nuclear energy were developed. The materials were inspired by a web-based Norwegian resource material[3] and reflected the findings of the ROSE-investigation (Sørensen, 2008) and the curriculum of physics/chemistry with its emphasis on connecting theory and experiments. Furthermore, the teaching materials also integrated the findings of Stadler, Benke & Duit (2001) about the different use of language and the idea of collaborative learning as well as the use of ICT-enhanced pupil learning. Story lines were produced in cooperation with one of the experienced teachers, Jens Grothe Møller (Vinding Skole) in three separate topics:

---

3   http://www.viten.no/

- Health and Radiation
- Nuclear Power Plants
- Nuclear Weapons

Each storyline posed a real life scenario and challenged students to come into the story as real world scientists and to make a scientifically informed case to decision makers and policymakers. For example, the story line of *Health and Radiation* posed the following scenario and asked pupils, with the guidance of their teacher, to make a video, rich in scientific knowledge, to provide answers to the range of questions posed.

> You often hear about nuclear radiation in the media. Sometimes you hear about the danger of emissions and accidents in connection with Nuclear Power Plants. Sometimes you hear about radiation used to cure cancer, and sometimes you hear about the long term effect of nuclear weapons used against Japan in the last days of The Second World War. How is this connected?
>
> In the next weeks you are going to work with the health aspects of radiation. How can radiation affect the body, and how can this kind of radiation be used in the hospital services? Imagine that you are going to establish a new local cancer hospital. But in this connection you have to be aware of certain aspects.
> - How thick must the shield be regarding radiation treatment?
> - How do you know when a body is free of a radioactive trace element?
> - What kind of radiation can be found and what kinds of particles are responsible for the different types of radiation?
> - What kind of radiation is most dangerous for the body?
> - How can radiation be used to cure cancer?
>
> You are going to produce a short video which presents and explains the above mentioned questions! Imagine that your video has been made by a group of scientists with the purpose of giving politicians a scientific background to make decisions. As you are a group of scientists, it is important that your video is based on results from research.

Different kinds of evaluation were involved to assess the effectiveness of this inquiry-oriented approach to the teaching of science principles and concepts. One of the tools used was teacher educators' observation of pupils' learning process. In another setting the experienced teacher and teacher educator evaluated the outcome, for example, videos made by pupils. The experienced teacher tested two classes with written questions – one class that had received traditional instruction and his own class which was involved in GIMMS. In addition, results from the final national examinations were compared with the evaluation of the video made by the pupils. The GIMMS approach will be continued with a new class due to start in the coming year.

The three different story lines developed for teaching nuclear energy appealed to different groups of pupils. The pupils' choice of story line reflected the findings of the ROSE investigations (Sørensen, 2008): for example, the story line *Nuclear Weapons* was chosen by 4 boys and no girls. *Health and Radiation* appealed to 11 girls and 3 boys and *Nuclear Power Plants* to 2 boys and 1 girl. The use of nuclear energy in power plants is the most commonly seen perspective and syllabus item in traditional science teaching material in Denmark, but it was the least popular story line selected in the class. Although the story lines were constructed to meet the demands of the curriculum, the possibility of working with a certain degree of freedom and choice appealed to most of the pupils.

The evaluation of the videos reflected different levels in the pupils' performance. Two groups with a total of 7 pupils, and of mixed gender (*Health and Radiation* and *Nuclear Power Plants*), achieved the best outcomes but with completely different video results. The pupils were able to combine theory and experiments in their explanations, which also reflected views from the media and society. Another mixed gender group with 4 students (*Health and Radiation*) was less convincing, but they were 'on their way' regarding their explanations. A group of 4 girls (*Health and Radiation*) demonstrated that they were able to explain using both theory and experiments and a group of 4 boys (*Nuclear Weapons*) almost reached the same level. In one group of 3 girls, the explanation in the video *Health and Radiation* only referred to descriptions directly from the textbook with no effort made to link theory and experiments.

Compared to the class that had received traditional instruction, the class involved in the GIMMS project received similar test results in the written test. However if the theme (decided by a lot) in the final, oral examination was about nuclear energy, there was a real possibility of better pupil performance as this group used a more detailed and precise language in relation to nuclear energy. They showed they understood the issues and were better able to communicate their findings.

Changes in practice

From the beginning of the project we focused on new and innovative ways of collaboration between the three different groups participating in the project: student teachers, experienced teachers and teacher educators. To start the project, we arranged a "Curriculum Workshop" (Lang, Couso, Elster, Klinger, Mooney Simmie & Szybek, 2007). The aim of the workshop was to develop agreements so that the three groups of participants worked together in an atmosphere of equality and respect. This democratic approach took into account a well-known hierarchy of positions within the groupings. The group work accomplished in the workshop continued afterwards – one of the groups was very successful in this regard. What normally in Denmark takes the form of a novice – expert relationship, where the experienced teacher serves as an evaluator of the student teacher, was here experienced as authentic cooperation. As one of the experienced teachers said:

> "For the first time in my long career as a mentor for student teachers, I was challenged to go deeper into technical and pedagogical issues. Until now our discussions had primarily been of practical issues".

Student teachers' process of writing reflective journals

Usually student teachers complete a reflective journal during their practice placement. Often they do not put much effort into this reflective practice journal. In many cases neither the experienced teachers nor teacher educators give meaningful personalized feedback on these reflective writings.

Because of the intense focus on the practice period and because the student teachers and experienced teachers were involved in collaborative planning of the GIMMS project, student teachers were required to write a reflective practice journal. As a result of this reflective writing and the feedback generated in discussion with the experienced teacher they found this reflective demand interesting and educational.

The double approach of the experienced teacher: the organization of learning in classroom-teaching *and* a focus on what is to be taught

Experienced teachers have professional autonomy and are responsible for planning, teaching and reflecting on their classes. Normally, experienced teachers have little or no opportunity to collaborate on these matters with teacher educators – or even with other experienced teachers at the school site. In Denmark, there is no significant tradition of teacher teamwork when planning the details of teaching or its evaluation.

In the project, the teacher educators aimed to support the experienced teachers in their planning and teaching during certain periods, and they collaborated in different ways such as jointly commenting on the main goals for a teaching period. Furthermore, the role of the teacher educators was to introduce relevant literature that would enhance understanding, for example, of the issues associated with gender. The project gave teacher educators the possibility of observing classes in case study schools and of sharing observations and reflections afterwards. Some of these classes were recorded on video in order to share observations and reflections. In addition, collaboration between experienced teachers and teacher educators included a range of curriculum making processes, including both short and long term planning, the development of classroom materials and student teacher and pupil evaluation.

Data collection and analysis

During the GIMMS project different kinds of data were collected, and due to the case study nature of the project, more qualitative than quantitative data were collected and analyzed. The following list includes the seven sources of data that gave the exploratory qualitative findings from the project:

a) Videos from the curriculum workshop, the seminar, and a lecture for pupils from 7th–10th grade, taught by a guest teacher invited by one of the experienced teachers.
b) Forms completed by experienced teachers and student teachers. Focal points were problems and anticipated difficulties and how they planned to solve them.
c) Questionnaires completed by experienced teachers.
d) Observations and reflections made when visiting schools.
e) Documentation of collaboration between student teachers and experienced teachers.
f) Reflective journals from student teachers.
g) Reflections on collaboration with experienced teachers when planning lessons and developing materials.
h) Observations and evaluations of the students' work with the new materials in school.

Results – changes of practice

In the beginning of the GIMMS Denmark project all groups focused on educational innovation and change by working with modified classroom organization. Change was mainly concerned with issues about dividing the class into small groups with one focus being on gender and another on skills and competences. Tasks or materials were not taken into consideration at that time.

The project made it possible for experienced teachers to become more involved with student teachers in discussions about educational innovation and change and aspects of teaching and learning, as they planned the lessons together. This was new, and as such innovative, and involved an accepted collaborative model of mentoring and reflection. The traditional mentoring relationship was changed from a novice-expert to a relationship of equality and mutual respect. In addition, it changed the situation. Generally, teachers have no time to engage in theoretical discussions, but tend to focus on general and practical issues and keep the actual teaching a private matter. Planning the students' practice period was the central part of the mentoring relationship, and the SMTTE model (Balle & Mølgård, 1997; Andersen, 2000) was used in the process of planning (Figure 3). The SMTTE model considers the cycle of learning as being made of a number of different aspects: 'Sammenhæng (Context), 'Mål' (Aim(s), 'Tegn' (Signs), 'Tiltag' (Action) and 'Evaluering' (Evaluation).

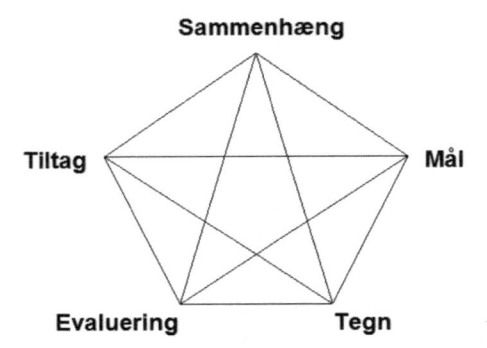

Figure 3: The SMTTE model considers: 'Sammenhæng (Context), 'Mål' (Aim(s), 'Tegn' (Signs), 'Tiltag' (Action) and 'Evaluering' (Evaluation).

Student teachers kept a reflective journal from their practice period and shared it with the GIMMS participants on the ICT platform. In the journal they focused on group organization, and they divided students on the basis of gender. They found it to be relatively easy working with the girls and less so with the boys.

As a result of the work between student teachers and experienced teachers, the former prepared their teaching in their school practice with a focus on gender and innovation. They taught mathematics as well as physics/chemistry, and in both subjects they focused on classroom organization of the learning environment. Regarding mathematics they made the groups "different" from what the pupils were used to. In fact, pupils were initially grouped according to skills – but after a week the student teachers decided to rearrange the groups again. In physics/chemistry boys and girls were taught separately, and student teachers concluded in their reflective journal:

> "It is ... easier to teach the girls ... and ... hard to motivate the boys – except for discussions on explosives ... and it's ... difficult to get the boys to behave (!)"

Some of the advantages for student teachers in GIMMS Denmark was the additional time to collaborate with experienced teachers. Although this gave the student teachers the opportunity to include their reflections and findings in essays and bachelor projects, the GIMMS project showed a need for change in Danish teacher education. It is important to have more precise instructions and opportunities at college to include school based projects and reflective journals in student teachers essays and assignments.

In the mentoring relationship between the experienced teachers and the teacher educators the focus was on developing teaching that would reflect innovative practices, for example, the findings of the gender-sensitivity of interest from the ROSE-investigation (Sørensen 2008). Different story lines for teaching Nuclear Energy were developed, and modules called *a. Health and Radiation, b. Nuclear Power Plants and c. Nuclear Weapons* were designed. The *innovation* agenda in GIMMS *Denmark* was to create a teaching and teacher education approach that focused on gender-sensitivity, student involvement and collaborative learning.

The reflective teacher

In the Maynard & Furlong (1995) studies the time consuming nature of authentic collaboration is crucial for the reflective mentor model. In the GIMMS project the experienced teachers were offered additional and 'specific' time to engage in the mentoring relationships with the student teachers and the teacher educators. Some of the student teachers were offered practice periods at the case study schools, but it was difficult to match all interested student teachers and experienced teachers in the project. Some student teachers joined the project without being able to participate directly during their time of study, but because they were interested in the focus of the project. They really intended to put work into the project – but for several reasons they dropped out again after a short period.

Some of the advantages for the participating student teachers were the mentoring relationship including extra time to collaborate with experienced teachers. GIMMS Denmark gave them the opportunity to include their reflections and findings in essays and bachelor/examination projects. Some of the findings from GIMMS show the need for more precise instructions and opportunities on how to use school based projects of this type and reflective journals in essays and assignments.

In Denmark, like in other countries, we strive to educate teachers for better development. What kind of knowledge do student teachers have to acquire – theoretical knowledge or practical knowledge? In Denmark they have to acquire both – but the question is how to combine the two kinds of knowledge in forming a career-entry qualified teacher?

The project showed how the experienced teacher and the student teacher can benefit from cooperative planning and reflection time. The experienced teachers were challenged to work more thoroughly with pedagogical and technical issues than they normally do because of reflections and discussions with student teachers. On the other hand, student teachers were challenged to combine practical experiences from the classroom with theoretical reflections. Reflective journals from teachers show that while they have to struggle developing their professional skills, the mentoring relationship might influence the experienced teacher's own professionalism in a positive way.

Teachers in the Danish lower secondary (comprehensive) school tend to focus on general teaching issues (child-centered) – rather than issues related to the subject matter knowledge. Because of that, they merely address problems concerning teaching in general – not the possibilities and barriers connected to teaching mathematics or sciences specifically. In addition, experienced teachers in Denmark tend to concentrate more on the hands-on approach rather than finding their support in theories – neither educational nor subject related theories. Experienced teachers in Denmark have no tradition of reading theoretical literature – neither educational nor subject related, and because of that they do not seek inspiration or help from the research literature to cope with, for example, gender problems. But collaboration in well structured research and development projects, like GIMMS Denmark, can make it clear for all involved, teachers across their professional lifespan and teacher educators, that practice and theory can be combined to develop new thinking which can be shared with both schools and teacher education institutions.

The international dimension of the GIMMS project has emphasized the different models of mentoring throughout Europe. While there were many points of agreement between the countries in GIMMS there were also many points of difference. However, the differences were rewarding in trying to seek an understanding of the complex nature of mentoring relationships, and they were a result of differences in teacher education, the view on research/developmental projects at universities/ university colleges, the different structures of school practice, curricula in schools etc. In addition, the international dimension gave us an insight not only in 'problems' across national borders but also in the cultural diversity in relation to gender, motivation and interest. GIMMS provided an opportunity to discuss the relationship between theory and practice in matters of innovation and change.

While it is relatively easy for researchers to disseminate the findings of the GIMMS project in teacher education and in-service courses for experienced teachers, it is much more difficult to implement changes at organizational and policy level. Nevertheless, it is necessary to address these organizational layers as, for example, different boards and committees, to have a long-lasting effect and to promote further learning, development and innovation in teaching, teacher education and in schools for educational innovation and change and gender inclusion.

Chapter 5:
Germany I: subject-related dialogic and democratic mentoring in biology teaching and teacher education

Doris Elster
Universität Bremen, Didaktik der Biologie, Bremen, Germany

Background and context

Teaching and learning to teach are complex professional competences. In the last decade mentoring for the support of teachers' professional development has been at the core of empirical studies (Fischer, v. Andel, Cain, Zarkovic-Alesic & v. Lakerfeld, 2008). Most of these studies intend to assist mentors in their mentoring focusing on the professional growth of student teachers, called trainee teachers, during their first years at school. Only few studies investigate the mentoring processes in the earlier phases of academic education (Elster, Albrecht, Hitzenberger, Kartusch, & Stadler, 2006; Elster, 2006; Ecker, 2005). The present study is an example of innovative high school education in which a school-based practical course for ongoing biology teachers and teacher continuing education will be analysed.

How does one become a teacher in Germany? There are two phases in teacher educational programmes: academic training at the university and school-based practical education. During the first phase, trainee teachers are engaged mainly in subject-related courses that are part of academic disciplinary learning. Although this first phase of academic teacher education lasts five years there are few opportunities for trainee teachers to gain practical classroom experience. Innovative courses are needed to connect academic education and practical school-based experiences. The first phase is completed with the first state examination or – as a result of the Bologna reform of Bachelor and Master programmes – with a master thesis. The second phase of teacher education is the practice-oriented preparatory service. In general this phase lasts one and a half to two years. During this initial phase of practical work young teachers are promoted and supported by experienced teachers.

In Austria, the teacher education program is very similar to the German one, but the second phase is shorter. In general it lasts one year.

In terms of mentoring, both phases support different research perspectives. The GIMMS Germany II study of Lang and Friege (see Chapter 9) focuses on the second phase.

In this German study I the author developed a combined practical and theoretical course for the first phase of biology teacher education. This practical course is an innovative model of school-university cooperation combining theoretical and practical elements of teacher education. In this course, trainee teachers (mentees), experienced teachers (mentor 1) and teacher educators (mentor 2) work together. The main goals of this cooperation on teaching and teacher education lie on two levels:

On the *individual level,* teacher education and professional development of trainee teachers as well as experienced teachers and teacher educators are expected. All participants in the practical course are "learners" and are seeking personal growth. That requires a learning environment of trust, mutual respect and ethical responsibility.

On the *system level,* mentoring within the practical course is "a strategy of change" that aims to introduce a new culture to schools as well as to the university. This educational innovation and change is a re-culturing process that has a focus on curriculum and teaching. That needs to be completed in a community that is characterized by mutual learning and flat, rather than hierarchical structures. The community of learners consisted of trainee teachers (mentees) and two mentors with different experience and background knowledge, who came from different institutions (school and university). Therefore, the author has called the mentoring process within the teams "dual mentoring".

Theoretical frame

The GIMMS Germany I study uses mentoring as a "strategy of innovation and change". Based on Hargreaves & Fullan (2000) teachers (or – in our study – trainee teachers) require mentors who guide them through developing the skills, abilities, dispositions and attitudes they need as professionals. What are the steps of professional development in the life of a teacher? Which changes are required during the professionalization process? Hargreaves et al., describe four "ages" or steps in developing teachers' professionalism, and find connections between these steps and the role of a mentor (Hargreaves & Fullan, 2000):

- In the *pre-professional age* teachers learn to teach by watching others. Mentors support the teachers by giving helpful advice to encourage them. Teachers without experience are modelling the mentor.
- In the *age of the autonomous professional,* teachers act as individualists. Teachers assume that they can manage themselves, and needing help is seen as incompetence in doing their jobs. Therefore, mentors are seldom contacted.
- The *age of the collegial professional* is characterized by increased complexity, knowledge explosions and new challenges. Collaborative learning cultures need continuous support from colleagues. Mentors are partners integrated in collegial teacher teams where mutual learning is promoted.
- The *post-modern professional age* is characterized by diversity and new partnerships beyond schools. Settings with school – university cooperation facilitate

mutual learning for personnel from different institutions. Mentors and mentees are all learners in this challenging process.

In GIMMS Germany I we focused on the post-modern professional age in a school–university partnership to challenge traditionally inherited practices and develop educational innovation and change. With respect of biology teacher education we established a new partnership between a university and schools, bearing in mind that mutual learning of a teacher educator, teachers and trainee teachers are core elements within the mentoring process. The goal was to establish a win-win-relationship of learning where mentees as well as mentors benefit.

Educational innovation and change

In GIMMS Germany I innovative educational practice in teaching and teacher education occurred on two levels:
- On the first level a new model of biology teacher education, a combined practical and theoretical course, was developed with "dual mentoring" as a core element. Trainee teachers, experienced teachers and teacher educators worked together in communities of practice (Wenger, McDermott & Snyder, 2002). The teams were characterized by a flat structure and a constructivist learning atmosphere in which reciprocal learning was supported. Team members learned by their collaborative work, joint development of materials and teaching units and collaborative reflection on classroom experiences. In this social and communicative setting, new ideas circulated and mentees and mentors shared their thoughts and knowledge.
- On the second level innovative educational practice occurred by developing a number of products (classroom materials). These innovative teaching materials supported contextual learning. Subject knowledge and competence became meaningful in a particular gender-sensitive context supporting social learning. The goal was that the learners (of the secondary level) learned by acting together. In discussions, debates and dialogue, understanding of subject topics in socio-scientific contexts were compared and confronted. This deliberative dialogical approach facilitated innovation and change through creating a 'public space' for communication and justification.

Dialogic mentoring

In GIMMS Germany I the author advocated for Colwell's (1998) considerations. He defines mentoring as a complex social correlation of people with different forms of experience, and as a *"process (correlation) within a mentor and a mentee (and within a group of mentees) in the atmosphere of trust, friendship, respect and taking care of each other in respect of pedagogical acting."* (Colwell 1998, p. 313)

The specific mentoring model used in GIMMS Germany I was based on the theoretical framework for *dialogic mentoring* according to Bokeno & Gantt (2000). This model characterizes successful mentoring as a 'dialogue': defined as a joint, constructive, critical-reflective, engaged and full valued discussion. It comprises: equality, empathy, real interest, reflectivity, mutual engagement and border crossing in communication. Mentoring in this point of view is not a directive, it is not a one-sided conversation led by the mentor, but a mutual process of communication at eye level. Mentors and mentees "construct together reality (and don't only reflect on it) … this is characterized by a joint inquiry … The basics of communication are critical questions and reflections … but not ready answers." (Bokeno & Gantt, 2000, p. 238)

To analyze the processes of mentors and mentees professional learning the author used a model developed by Niggli (2003, 2004) focusing on three different levels of the mentor-mentee relationship. This model of "Three-Level-Mentoring" included the levels of action, communication and personal development (Figure 1).

The first level, the level of action, is concerned with the visible performance of one's practical actions. These actions are included in the discussion between the mentor and the mentee. They exchange information about *knowing how* to teach in a very concrete way. Skills of teaching or the teacher's behaviour are kept in view. The core of the communication of mentees and mentors is a direct feedback on action.

The second level, the level of communication, is concerned with the background knowledge behind the visible actions. This knowledge integrates scientific and educational theories, arguing from the standpoint of *knowing that* instead of *knowing how*. The personal development of the mentees is based on the didactical or educational background knowledge. The mentor-mentee communication is a reflective meta-communication about the teaching and learning situation.

The third level of mentoring, the level of personal development, is concerned with the *professional self* of the mentee, his or her self-esteem and consciousness of success and failure. The mentor-mentee communication is a reflective meta-communication about personal issues. The mentees' personal development is promoted by

Model of Three-Level-Mentoring

Action		Communication	Personal development
Professional self/ values/ motives	Mentoring Level 3	Reflective meta-communication: personal issues	Clarifying knowledge about the professional self
Theoretical and practical background	Mentoring Level 2	Reflective meta-communication: background knowledge	Differentiating teaching and learning background knowledge
Practical acting	Mentoring Level 1	Feedback on acting	Optimizing skills

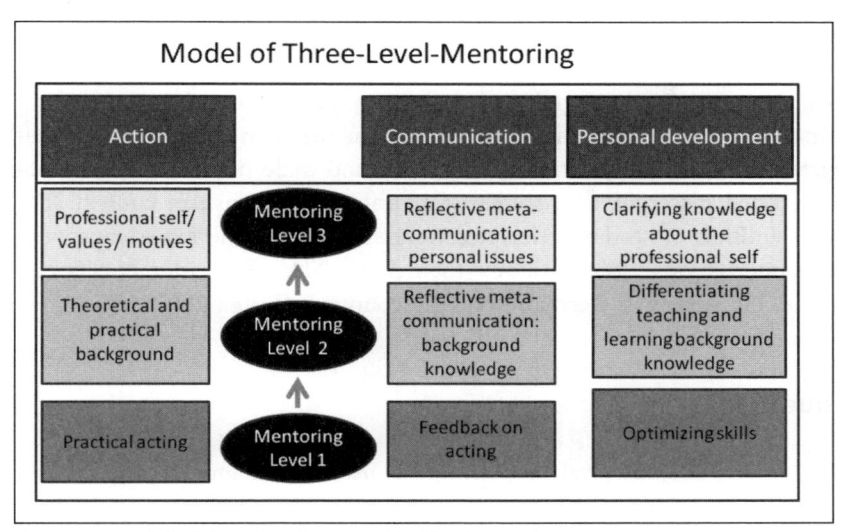

Figure 1: Model of Three-Level-Mentoring (Niggli, 2003).

clarification of knowledge about the professional self. Table 1 gives an example of the communication in a mentoring process and how to analyse the three different levels of communication.

Table 1: The mentor-mentee communication

A trainee teacher (mentee) wanted to promote inquiry-based learning in his classroom. The pupils had to plan and conduct experiments to investigate the food preference of soil animals. But the pupils' results were very frustrating since they had problems on different levels. The frustrated trainee teacher decided that he would not perform such lectures in the future. The trainee teacher's experiences can be reflected on three different levels during the mentoring process:
- *First level* – level of concrete practical action: Mentor and mentee discuss if the task for the pupils was clearly formulated or if the task was complex enough to promote inquiry-based learning.
- *Second level* – level of theoretical and practical background: Here the mentor-mentee communication deals, for example, with questions about the pre-knowledge and the abilities of the pupils, and whether they have enough background knowledge to solve the inquiry-based task.
- *Third level* – level of the professional self: The mentor-mentee communication can focus on the trainee teacher's feeling of frustration, on his values and teaching goals or on possible strategies to deal with a lack of success.

Gender awareness

In GIMMS Germany I we explored two aspects of gender-sensitive pedagogical practices. Firstly, we focused on the processes within the mentoring process itself and investigated the different questioning of female and male trainee teachers. Secondly, we developed materials in a student-relevant, gender-sensitive context. The basis for the definition of student relevance we gleaned from the experiences and results of the international ROSE survey (Schreiner, 2006: Schreiner & Sjöberg, 2006) and the ROSE survey in German-speaking countries (Elster, 2007).

Research questions

Niggli's (2003) model of three-level-mentoring was used as a helpful tool to analyse the mentoring process (Table 1). For example, on level 1 the mentee received a precise response or feedback on his or her classroom behaviour and skills; on level 2 mentor and mentee talked about explanations or reflected on teaching knowledge on a meta-level; and on level 3 the situation was reconstructed, aiming at reflection on attitudes, dispositions and further decisions regarding teaching.

The following research questions were at the core of the GIMMS Germany I investigation:
1) Which levels of mentoring can be identified by analyzing the communication of mentors (science educators and teachers) and mentees?
2) What are the themes of the communication between female and male students and their mentors?

Additionally, based on the theory of dialogic mentoring (Bokeno and Gantt, 2000), the author wanted to know:
3) Could the ideal of a dialogic mentoring process be realized within GIMMS?
4) What are the products (gender-sensitive materials) of the mentoring process?

Dual mentoring process

The practical GIMMS Germany I course was developed at the University of Kiel, Germany in the years 2007 and 2008 and disseminated at the University of Vienna in Austria in 2008/2009. In this practical course teams of six to eight trainee teachers (mentees), a teacher educator from the university (mentor 1), and an experienced teacher (mentor 2) worked together as a community of practice (Wenger et al., 2002). Thus, two mentors with different experience and background knowledge, who came from different institutions (school and university), worked together. The author called the process within the teams "dual mentoring". The teams worked together for half a year. Together, they developed gender-sensitive materials and units, tested these materials in the classroom, and reflected on classroom experiences and the mentoring process.

The team structure was characterized by a flat, rather than a hierarchical, format. The goal was the establishment of a win–win relation for all participants. The *piloting courses 1 and 2* at the University of Kiel were conducted with advanced students. The *dissemination courses* at the University of Vienna were conducted with trainee teachers at the beginning of their academic training. Figure 2 gives an overview of the timeline of the project.

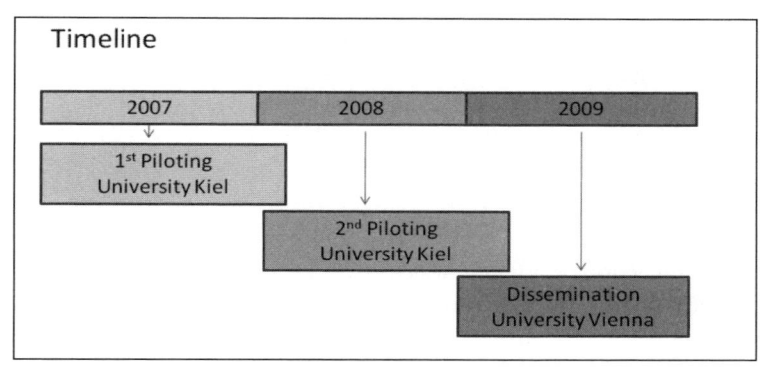

Figure 2: Timeline of the GIMMS Germany I project.

Piloting of GIMMS Germany I

Piloting at the University of Kiel was conducted within the "Interdisciplinary Practical Course" for advanced trainee teachers at the Leibniz Institute of Science Education, Department of Biology Education. This was an optional course within the academic program for current biology teachers at the University of Kiel. The course was offered in spring 2007 and in autumn 2007/ 2008. The target group of the course were advanced students (eight participating students in 2007, six participating students in 2008). The author herself acted as a science educator (mentor 1) and worked with an experienced teacher (mentor 2). In 2007 the participating school mentor was a teacher at the Gymnasium Heide, in 2008 mentor 2 was a teacher in the Gymnasium Marne. Both schools were located in the German federal state of Schleswig-Holstein.

The mentors' and mentees' task was the joint development of gender-sensitive materials, testing these materials in the organizational frame and context of a school project, and reflecting on their classroom experiences together. Mentees and mentors were asked to keep a reflective journal during the whole process. They noted not only what happened and what was observed during the meetings, but also they were asked to record their tentative hypothesis or their development of new understanding. The reflective writing was to provide mentees as well as mentors with a systematic approach to their development as reflective, critical, and constructive learners. The reflective journal was to assist the reflective process and was structured as a personal learning journey, tracking and documenting an evolving understanding of teaching, learning and innovation (Altrichter & Posch, 2003).

The reflective journals of mentors (N=4) and mentees (N=14) were analyzed according to the paradigm of qualitative content analysis (Mayring, 1998). The juxtaposition of the views of mentees and their mentors gave an insight into the complexity of the mentoring process. Additionally, photo protocols of each school project allowed to triangulate the concrete teaching and learning processes: mentees, mentors and students were asked to comment on the different teaching and learning situations by analyzing the photo protocols.

Dissemination

Dissemination at the University of Vienna was conducted within the "Pedagogical Practical Course" at the Austrian Educational Competence Centre Biology. This was an obligatory course within the academic training program of current biology teachers at the University of Vienna. The course was conducted in two semesters, in spring and in autumn 2009. In each semester, six mentoring teams participated in the project. In total, 152 trainee teachers, six experienced school teachers (mentors 2) and one science educator (mentor 1) participated in the dissemination phase of the project.

For the data collection the author used reflective journals and interviews with six selected trainee teachers and the six mentors 2 and a questionnaire. The questionnaire contained eight sections:
1. Personal expectations on the practical course (5 open questions)
2. Themes discussed at school (9 items, 5-point-Likert-scales)
3. Themes discussed at the university (6 items, 5-point-Likert-scales)
4. Cooperation on the practical course (5 items, 4-point-Likert-scales)
5. Support (15 items, 4-point-Likert-scales)
6. Working at school (6 items, 4-point-Likert-scales)
7. Gender differences (11 items, 4-point-Likert-scales)
8. General questions regarding teacher education (5 open questions)

In addition, questions were asked on demographic data (gender, number of semesters, subsidiary subject, number of students in the team, time of performed lessons at school).

Outcomes from the project

An interdisciplinary model of subject-related teacher education

On the system level, the "Interdisciplinary Practical Course" constituted an innovative model of teaching and teacher education. It offered student teachers the opportunity to gather practical experiences of teaching and curriculum development at an early stage of their academic training. Additionally, within the course, the communication of stakeholders of two different institutions (experienced school teachers and science educators from the university) was promoted successfully. The course allowed collaborative learning of trainee teachers, teachers and science educators across borders. Mentors as well as mentees grew in the process of innovation and change and there were benefits on both sides.

Gender-sensitive materials

Further outcomes of the GIMMS project were the generation of a number of materials and units that were gender-sensitive regarding their contexts. The GIMMS teams suggested that the context could be a powerful vehicle for acquiring knowledge and competence. Based on the results of the ROSE study in Germany and Austria (Elster, 2007; Schreiner, 2006) the contexts chosen need to be of social-scientific interest and relevant for girls as well as for boys.

Therefore, in the *GIMMS piloting phase* we chose themes in the context of medicine, the human body, ethical judgement, and moral decision-making:
1. *"Pretty, healthy, fit – a life show"* was a teaching and learning unit for 9-10 graders. Eight persons were invited to participate in a TV debate about fitness, health, and developing strong and good-looking bodies. Nutrition and eating disorders as well as fitness studios and beauty queens were themes that were discussed.
2. *"Lara is pregnant"* was a teaching and learning unit for 8 graders. Lara is 16 years old and in the 9th week of an unplanned pregnancy. She and her classmate Jonas, the father of the child, do not know if they should have the baby or not. They are not quite sure about the consequences of whatever decision they make. Therefore, they need to ask for help with this decision.

Both teaching units consisted of a series of lessons where a semi-real dilemma was used to stimulate a moral cognitive process in the students. Therefore, the KMDD (Konstanz Moral Dilemma Discussion) by Lind (1978) was used as a theoretical frame. This method established merits for effective moral learning and democratic learning. The teacher puts the students into a semi-real situation and confronts them with a controversial discussion. That creates emotions and social reactions that need to be taken into account. To deal with this situation, the students must activate and develop their moral and democratic competences.

The materials of the *GIMMS dissemination phase* differed from that of the piloting phase, especially in their length. They were relatively short (two-hour) teaching units with domain-specific content. As in the piloting phase, the contextualization of the subject content was also a core element.

Here are two examples of the developed material:
1. *"How to plant strawberries?"* was a teaching and learning unit for 5 graders. Children wanted to grow delicious strawberries in their school garden. Therefore, they discuss in the unit the reproduction procedure of the species that they have decided to plant in a democratic way.
2. *"Golden Rice for India?"* focused on aspects and procedures of Green Genetic Technology. Two girls with different cultural backgrounds (Sarah lives in Bremen, Germany, and Saranja lives in India) discussed the benefits and the dangers of sowing vitamin A-enriched rice ("Golden rice") in an Indian rural region. The teaching unit deals with the procedure of genetic changes and the unpredictable impact of genetically modified plants. The unit has a social-cultural context including historical elements such as the "Green Revolution" in India.

Findings and discussion

Which levels of mentoring could be identified by analyzing the communication of mentors (science educators and teachers) and mentees?
Based on the results of the interview survey and the reflective journals (piloting: 14 trainee teachers; dissemination: 152 trainee teachers) the communication between mentor 1 (science educator) and mentees (trainee teachers) was mostly at level 2 (reflective meta-communication about teaching, learning and curricular matters). Mentor 1 and mentees seldom discussed at level 1 (direct feedback on actions in the classroom) or level 3 (personal coaching and development). School teachers (mentor 2) and mentees (trainee teachers) communicated more at level 1 than mentor 1 (science educator) and mentees. Discussions at level 3 were rare too. It was assumed that this was caused by a lack of time. An overview of the levels of mentoring gives Figure 3.

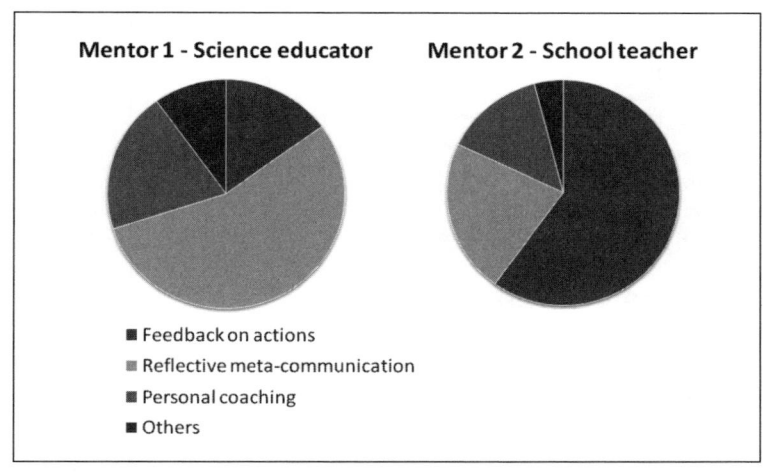

Figure 3: Levels of communication between mentors and mentees. Data based on the interview survey (6 mentees, 6 mentors) and the analysis of reflective journals (piloting: 14, dissemination: 152).

What were the themes of communication of female and male students and their mentors?

The data of the questionnaire survey (section 2: "Themes that were discussed at school with mentor 2"; section 3: "Themes that were discussed at the university with mentor 1") confirmed the findings of the qualitative analysis of the reflective journals, namely that mentoring discussions were mainly at level 1 (with mentor 2) and at level 2 (with mentor1). Examples of items coded as "level 1" were "lesson planning", "exchange of materials", "classroom experiences". Examples of "level 2 items" were "individualization", "motivation and interest", "curriculum", "parents' expectations", "choice of methods". Discussions at level 3 were rare. Items coded at level 3 were "clarifying the social status" and "clarifying the teacher's role". Figure 4 gives an overview of the questionnaire analysis (section 2 and 3; 5-point Likert scales; seldom – often). The findings indicate gender differences: in general female trainee teachers communicated more with the mentors than male students. There were high significant differences in the frequency of the discussion about "classroom experiences", "motivation and interest" and "clarifying the social status".

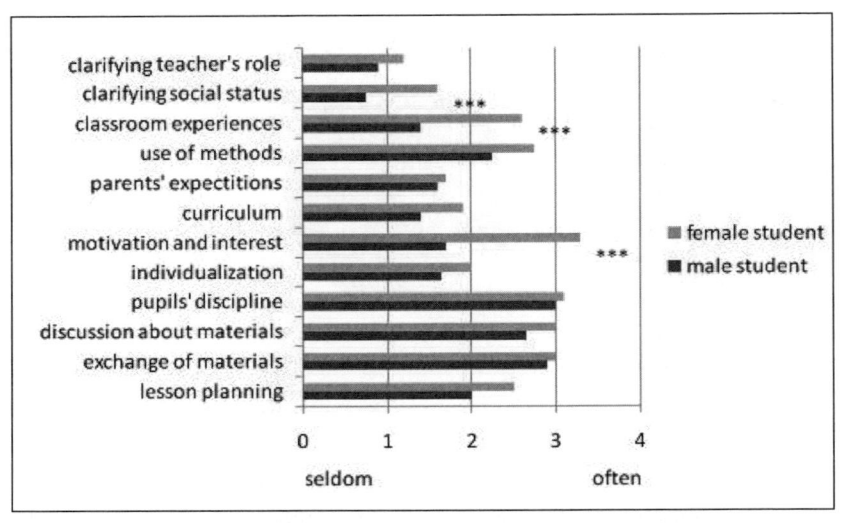

Figure 4: Themes of mentoring communication. Findings of the questionnaire survey (152 teachers students; ***p<0.001)

How far could the ideal of "Dialogic Mentoring" be realized?

Based on the results of the pilot phase (reflective journals of students and mentors), mentors and mentees described their relationship as one of mutual respect, and the atmosphere as constructive cooperation. The mentoring teams were characterized by a flat, rather than a hierarchical structure. Mentors and mentees worked together in an atmosphere of trust and mutual judgment for the development of gender-sensitive materials. The students defined the criteria of a "good mentor" as followed (results of the analysis of the reflective journals):

> *Competence:* The mentors have theoretical and practical competence in their domains. The cooperation between a pre-service educator (mentor 1) and a school mentor (mentor 2) is fruitful as they are competent partners in different areas.
>
> *Reflection:* Good mentors can accept feedback and critical remarks. They are able to implement personal development and change.
>
> *Standards and values:* Good mentors are trustworthy and punctual (as expected by the students).
>
> *Personality:* Good mentors are flexible, creative, resolute, and have foresight. They have a "feeling" for the needs of the mentees. They are "mature" persons with rich experiences.
>
> Student S: "Good mentors give support, encouragement, and security to the mentees, and develop their self-confidence." (ST_S_rj_12)

School mentor 1: "During the GIMMS-project the students and I had such a (dialogic) mentoring relationship, with an intention to guide the learning process of the mentees as they developed and experienced the "Lara" materials and evaluated the practical unit together ... The most critical point for me was to wait until the students came up with ideas and questions and not to act as someone who told them how to behave in the classroom. ... The students brought a lot of new ideas (e.g. a talk show as a new methodical approach.) " (Ms_S_rj_6)

Mentor 2: "In my opinion the project was successful for both mentors and mentees. The mentees were able to gain concrete experiences in teaching and to reflect on their actions using the support of the mentors. I appreciated the opportunity to engage in communication with the current teachers. The cooperation also gave new impulses to my own planning and delivery of lessons and instructions. Since then, I have been pleased to recognize that the theoretical background coincides with my personal conviction of the understanding of a mentoring correlation." (Ms_ J_ rj_ 3)

In the mentors' opinion, the relationship between mentees and mentors was one of mutual respect and was developed in a constructive atmosphere, but also included critical and self-critical reflection about experiences in the classroom. Innovation and change was realised by mentors in GIMMS Germany I when they engaged in a deliberative discourse and "learned" from mentees about the selection of relevant contexts in a project that involved joint inquiry, communication across borders and material development. Mentees reported that professional development was achieved by optimizing skills, differentiating background knowledge of teaching, curriculum and learning, and by clarifying knowledge about themselves and their ability to work as teachers. In this way educational innovation and change in GIMMS Germany I challenged traditional approaches to teaching and teacher continuing education through dialogic mentoring as a deliberative discourse.

Chapter 6:
Austria: video analysis for innovation, mentoring and inclusion

Helga Stadler and Susanne Neumann
Department of Physics Education, University of Vienna, Austria

Teacher education in Austria takes place at teacher education and training colleges and at universities, such as the University of Vienna. It all depends on the school type in which student teachers aspire to teach (Posch, 2010)[1]. Teachers who want to teach at grammar schools must have a university degree (Master's degree). The teacher education programme at universities consists of three parts: education in two subjects (such as physics and mathematics), courses in pedagogics and courses in didactics of each subject.[2] After graduating, the future student teachers have to complete one year of teaching at a school, during which they are supervised by an experienced mentor. Practical teaching experience in a school does not take up a large part of the curriculum at the university. During the first part of their study, student teachers observe lessons and then at a later stage they have to teach some lessons on their own. In that way student teachers should get a realistic idea of what it means to be a teacher.

Background and context of GIMMS Austria

We teach student teachers who aspire to be physics teachers at the University of Vienna. With regard to the topic of physics education, there are not many courses which are obligatory for student teachers. As a whole, student teachers of physics only have to take courses in science education which add up to 19 ECTS points (European credit transfer points) in their university education and training programme. In total students have to have 120 ECTS points in each of the two subjects. The primary focus of these 19 ECTS points is on the didactics of experimental physics, where the student teachers learn how to do demonstration experiments. Only about 5 ECTS points are given for courses that deal with topics of general physics education. Of course there are some students who also take additional non-obligatory courses or focus on science education when they have to choose elective courses.

One of the courses which focuses on physics education is called "Teaching Physics: A Practical Approach" which is designed for student teachers who are already par-

---

[1] http://www.bmukk.gv.at/schulen/bw/ueberblick/bildungswege.xml; a critical review on teacher education in Austria see P. Posch, 2010.
[2] http://lehrerinnenbildung.univie.ac.at/

ticipating in the student teaching phase of their education. The main objective of the course is to support these students in a direct way by developing and discussing lesson concepts, by personal coaching and by giving them an insight into legal issues such as school curricula and employment situations. A second type of support is more indirect. Current results of research in science education are presented which should assist students' reflections on their teaching methods. These topics include students' conceptions, PISA findings, gender issues, teaching methods and innovative approaches such as e-learning, science lab or the role of experiments.

The theoretical background of the design for our courses is an action research model developed in Austrian in-service training courses (Altrichter & Posch, 2009; Kühnelt & Stadler, 1997, 2008). For the purpose of reflecting and interrogating their own practice, student teachers learn to apply different methods of action research, e.g. how to develop and analyse lessons or plan and conduct interviews with their pupils. One of the main tools are videos, which they learn how to use. It is this latter part of the programme that was studied systematically in GIMMS Austria as a new and innovative teaching and teacher education approach to support pupil learning and inclusion.

The use of video in teacher education

Dealing with everyday processes within a public space for democratic discourse is at the core of our understanding of educational innovation and change. Within GIMMS Austria this understanding was shared with other colleagues from six different European countries. The task of science educators, as defined by our key principles, is to provide supportive perspectives through feedback and suggestions for teachers' own investigations, and ideas for their conceptual work. We developed special methods for using videos in order to provide different perspectives and to support professional discussions with mentors and colleagues about teaching and learning processes (Stadler, 2003, 2005).

The student teachers videotaped one lesson. They were provided with some questions to help them reflect on their videos. For example: What sort of questions do I use in my lessons? Who answers these questions? Do you notice differences in the questions of boys and girls? They were also encouraged to develop their own questions. As a home-assignment, they had to prepare a presentation for a seminar. For this purpose, they chose three scenes (as video clips) which seemed to be most interesting and in accordance with their questions. In the seminar, they showed these scenes to their colleagues and discussed them. As a second part of this homework they made a transcript of these video clips and developed a written account of what they learned from this shared meaning-making process.

Videos proved to be an excellent tool to initiate professional talk about learning and teaching. All student teachers reported that it was important for them to share their work and to get to know their colleagues' work in class, to discuss it and to gain new ideas and perspectives.

Concept of the course "Teaching Physics: A Practical Approach"

The course, entitled *Teaching Physics: A Practical Approach* was first developed in 1993 by one of the co-authors (Helga Stadler) and its concept has been somewhat changed in the past few years. The second author, Susanne Neumann, had worked as a physics teacher for some years, before she joined the course as a lecturer in 2004. She changed the style of the course in some ways to better connect with current developments that are taking place in the school environment. Through her background student teachers also got the opportunity, through this coursework, to identify with problems young teachers have when starting their school career.

The basic structure of the course combined theoretical input of physics education topics and practical assignments for students. It was this combined approach which offered an opportunity to the student teachers to integrate theory with practice, to use theory in their practice and to develop research findings at the school. The student teachers had to take part in a variety of approaches including discussion, researching a rich variety of topics, including gender which was relevant to GIMMS Austria, presenting findings, analysing video clips of their lessons and conducting evaluations (Table 1). The course was designed to give student teachers the tools and the language to discuss their physics teaching and to continue to improve it for the purpose of pupil learning and inclusion.

Table 1: Variety of GIMMS Austria approaches used to integrate theory with practice.

Topic	Description of contents and methods	Assignment for students
Good Practice	Input and discussion about conditions that determine the quality of physics lessons.	Students have to find examples of good practice in their observed lessons and present them during the course.
Innovations in Schools	The conditions students meet in different schools are discussed. Innovative projects in school development concerning science (e.g. science lab, laptop classes) are presented.	Students themselves get background information about their schools and compare them to others.
Lesson Concept I	A specific field of physics is chosen to develop a lesson concept that the students can try out in school. In the last years we focused on the topics: energy, heat, electrodynamics and relativity.	Students have to do background research on the subject as well as on didactical issues about this field of physics. They have to prepare parts of the lessons in group work.
Lesson Concept II	Another field of physics is chosen to develop a lesson concept that the students can try out in school.	Students have to do background research on the subject as well as on didactical issues about this field of physics. They have to prepare parts of the lessons in group work.
Students' Conceptions	Based on professional articles, students get to know about children's ideas in science and how this should influence their teaching paradigm. As this part is extremely motivating and able to change students' views, quite a lot of time is dedicated to it.	Students have to read articles on a specific topic of children's ideas, do research on their subject and present the major ideas. Then they have to plan and conduct interviews to find out if these ideas are also present in adults' minds.
Classroom Methods	A theoretical input about classroom methods is offered, accompanied by a group discussion about the applicability in every-day school life. Videos of physics lessons help to identify pros and cons of certain methods.	Students have to reflect on their teaching (or on their experienced physics lessons) and analyze videos in group discussions.
Pupils' Interests	The major studies about pupils' interests are presented to make student teachers aware of how essential a well-reflected choice of context is.	Students have to find subtopics which could be of high interest to their future pupils.
PISA	International studies about pupils' achievement in sciences are presented.	Students discuss the results of these studies and reflect on possible reasons for Austria's rather poor results.
Gender Issues	Due to the huge gap between boys' and girls' interests and achievements in Austrian physics lessons the discussion of gender issues is a vital part of this course. Students are confronted with results of these studies and discuss measurements that can be taken.	Most students are not aware of the role that gender issues play in science education. The discussion of these topics is always very lively and fruitful.
Pupil Assessment	First we focus on different ways of composing written exams, then various methods of pupil assessment are presented focusing on alternative ways to assess and grade pupils.	Students compare various assessment methods and discuss assignments and problems encountered in tests from different teachers.
Video Analysis	The analysis of videos should help the students to examine their own teaching. Videography is a very effective way to confront students with their strengths and weaknesses.	Students have to record their own lessons and present some scenes during the course. The scenes are analyzed and commented in group discussions.
Methods of Evaluation	The importance of evaluation in the class room is made plausible and different models of evaluation are presented.	Students have to evaluate this university course using different methods of evaluation and reflect on the applicability for school life.

Impact on the student teachers' attitudes towards physics education

Research shows that usually courses in physics education fail to change attitudes towards physics education and also fail to change student teachers' professional beliefs. In the following we are concentrating on the question whether or not our course concept in GIMMS Austria, was able to make a change in practice and a change in mind-set and beliefs.

The guiding key question for GIMMS Austria was as follows: Do students' attitudes towards physics education change due to this university course and if so, in what way?

Research studies in our faculty showed that student teachers had very different opinions about physics education. On the one hand these perceptions appeared to strongly depend on the teachers they themselves had and their relationship to these teachers. On the other hand some of them never had the chance to explore the idea of what the meaning of "physics education" or "didactics" was. The reason for not having a chance to explore these ideas was that the student teachers could choose between very different lectures and seminars, some of them held by teachers talking about, and exploring, hands-on experiments and more of them held by theoretical physicists, discussing the theory of relativity. So the question for us was: how do student teachers think about physics education after they have attended our course?

As already mentioned the concept of the course was slightly changed in the year 2004. That is the reason why GIMMS Austria focused on research and development for the courses given since 2004. The data we used for our research was based firstly on the reflections on the videos students took from their own lessons. Secondly we also used written feedback comments made by the student teachers. We did not use questionnaires for statistical data analysis. So we clustered the main issues mentioned by the student teachers according to the following issues which we will consider in turn: has science education opened up new dimensions to the student teachers, how eager are student teachers to learn more about science education topics, were the methods used in the course suitable, what are the student teachers perceptions of courses for the aspiring physics teacher, the role of the co-lecturer and any additional comments.

Science education has opened up new dimensions to student teachers

Many students commented on the course as something that made them aware of a whole new field of relevance. This was a most remarkable finding in that we had not anticipated in the portfolios. It appeared quite often among the feedback comments. We assume that some of the student teachers focused on the subject-related courses in their curriculum and have encountered science education for the first time in our course. Here are some quotations we found to illustrate this insight:

"Until now, I've only dealt with subject-related aspects, an understanding of a whole new dimension has been given to me."

"Making aware, that teaching physics does not mean the same as teaching the contents of the textbook. Before this course, I thought that I had to discuss every topic from the textbook. This course gave me the self-confidence to plan innovative and independent lessons."

"The importance of didactics became clear."

"I appreciate it a lot that this course has addressed topics that have not come up in most of the other courses, e.g. how to assess pupils, what rights and duties do I have as a teacher, what influences the quality of lessons."

Student teachers are eager to learn more about science education topics

Among the many positive comments were a lot of comments that expressed deep interest in science education topics. Students appreciated the articles and books on science education that were presented in the course, some of them asked for even more information on these issues. Although most of the students still were not close to their university degree, they wanted to get an insight into in-service education and training. Some of them even asked during the course, if it was possible to attend in-service seminars outside of their formal curriculum. This can be seen as a sign of their interest in further education and training and the motivation to co-operate in science education projects. Again, some quotations we found that illustrate this:

"Very interesting topics, such as PISA, pupil assessment and gender."

"Interesting: what pupils consider a good physics lesson."

"It was great that articles and documents that were interesting to us were uploaded onto the internet platform, even if there was not enough time to discuss them in the course."

"It was extremely exciting for me to read the articles about pupils' ideas. As it gets harder and harder to connect with the ideas and thoughts of pupils, it is very important to concern yourself with those kinds of articles."

The methods used in the course were suitable

In our course we tried to focus on methods that were not only suitable for a university course but would be of use for the student teachers in their teaching job as well, either because the methods trained their abilities to present and communicate or because these methods will also be applicable at school. The most important methods were group discussion, presentations, e-learning platform, portfolio as an assessment method and different feedback methods. The following comments convey the value the student teachers placed in these methods:

> "It was good to present our results in front of the other students – after all we have to do this as teachers as well."

> "Interesting new methods."

> "I liked the Moodle platform. It was very convenient to have a central source of information for personal reading."

> "I think it was great that we had to do different assignments during the semester, e.g. presentations, interviews, video analyses."

Students criticise the small contingent of courses in science education

Due to the mainly positive experiences students associated with the course, a lot of their comments referred to the need for granting physics education more attention and more compulsory courses in the curriculum for physics teachers. Many students expressed the idea that they felt unprepared to start their teaching jobs, although they have followed a lot of subject-related university courses. Their comments give a deeper insight into this:

> "I would have preferred to have had more time to deal with the objectives of this seminar."

> "This was the first university course related to pedagogics which I did not find completely unsuitable and unnecessary. Finally, the actual process of teaching which is important for our future was discussed in detail."

> "More lessons like this, please, so that more lesson concepts (e.g. mechanics) can be discussed."

Additional comments

Although it was recommended that student teachers attend this course when doing the practical teaching phase, some of the participants took the course at another time. Therefore the question arose if the course made sense to this latter cohort of student teachers. We assumed that the importance of science education was mainly seen when students were fully involved in the life of the school. Nevertheless, the comments of the students who were not taking part in their teaching phase in the same semester as the course did not substantially differ from comments made by the other students.

We also expected another difference to occur: As the course had only been compulsory for the last two semesters we assumed that the feedback comments would get more critical. Whereas there were only 7-8 students in our course during the optional phase who we expected to be especially interested in science education topics, their number doubled when the course became compulsory. We implicitly understood that there were going to be some students in the course that were not open to science education topics right away. However, analysis of the portfolios showed that the most positive feedback came from last semester's course. Those comments especially revealed a strong development of the student teachers' attitudes, from a subject-related focus to a stance which we termed "open-to-science-education minds".

The role of the co-lecturer

Another question was of upmost interest for us: as the course was organised and held by two lecturers, Helga Stadler as a university lecturer and Susanne Neumann as a young teacher, it was interesting for us to see if our expectations were met. The main purpose of two lecturers with different backgrounds was the wish to fulfil two different roles. The experience in science education research as well as the practical approach seen by a teacher should encourage the students to see science education as a field of study that is of direct use for every-day teaching. It was also important for us to show the students that the participation in ongoing projects, e.g. EU-projects, such as GIMMS Austria, in cooperation with schools and the university and the work as a teacher go well together. For this purpose Susanne Neumann seemed to be a good role-model. When it came to preparing lesson concepts, the view of a science education expert was equally important as the opinion of a teacher who decided how realistic the chosen learning objectives were. Finally, the direct contact to schools was vital, as there was a "direct line" to the innovation and development that are currently taking place in the school environment.

Consulting the portfolios and feedback comments, there was only one negative comment which criticised the waste of resources to have two lecturers in a course with only eight university students. The rest of the comments relating to the lecturers showed that the role of both of them seemed quite clear and comprehensible to the students. It was pointed out that the ideas and practical advice coming from the teacher's side as well as the suggestions relating to the lesson concepts and comments on the video analyses were highly appreciated. The student teachers found

GIMMS Austria gave them a 'good insight into the working life' of a teacher and helped them in practical ways develop new and innovative forms of teaching for pupil learning and inclusion.

Chapter 7:
Spain: mentoring for innovative science education – teaching for achieving scientific competence with student teachers and mentor teachers

Digna Couso and Roser Pintó
Universitat Autònoma de Barcelona, Spain

In PISA scientific competence as the main goal of science education was not achieved for Spanish pupils. One reason behind this lack of success is that experienced in-service teachers have conceptual difficulties regarding the PISA competency framework, and as a consequence, do not know how to teach for the achievement of scientific competence. This is particularly the case regarding the sub-competence "using scientific evidence". In GIMMS Spain, we addressed this problem for both in-service and pre-service teachers by proposing a new mentor-mentee model with a focus on the use of this competence in the science classroom. The idea was that both mentors and mentees, working with teacher educators/researchers, learned from each other and improved their understandings.

First results indicate that, despite shortcomings of the initial tasks designed by both mentors and mentees, their interactions and work within the project have improved their understandings, measured in terms of quality of the designed tasks, when compared with previous results. The analysis also offers suggestions for the refinement of the mentor-mentee-teacher educator/researcher framework to support educational innovation and change in teaching and teacher education. An instrument was also designed for analysis of the quality of teaching tasks to develop key competences in this framework such as argumentation, using data critically and contextualisation.

Background and context

In Spain, as in some other OECD countries, students' results in PISA (OECD, 2004; 2007) have not been satisfactory. Spanish results in PISA 2006 show worrying outcomes as summarised in the following (Varela & Pérez de Landazabal, 2009):

> Spanish students' results on scientific competence are significantly below the mean of OECD countries (488, being the mean 500).

> The worst results on "knowledge of science" (with a mean of 490) are those regarding content on physical systems (477), while content of life systems (498) or earth systems (489) is much closer to the OECD mean.

> Knowledge about science is also worryingly below the OECD mean (489), in particular regarding those aspects related with understanding of scientific investigations.

One of the reasons argued for these results, among others, is the challenging distance between the described "scientific competencies" that PISA assessment tasks demand and the sort of competences that are demanded of students in the average Spanish science classroom. In this sense, the traditional science curricula, generally compartmentalized, un-problematical and strongly focused on knowledge of science – neglecting skills, knowledge about science and socio-scientific context – is now under scrutiny more than ever.

The situation has lead to recent changes in the national curriculum in Catalonia (Department d'Educació, 2007). The new Catalan curriculum is both standards-based and focused on the idea of teaching and learning for the achievement of *basic competences* (according to the well-known document DeSeCo (OECD, 2005a)), including specifically *scientific competence*. This new orientation of the curriculum implies a deep educational change, from the acquisition of knowledge to the capacity to use knowledge (Perrenoud, 2008). In this new curriculum, scientific competence is defined using a PISA inspired definition (OECD, 2004).

Despite the difficulties foreseen for such a curriculum change, in-service teachers who are now implementing this new "competence-based" curriculum have not received any special support regarding the teaching of science for the achievement of scientific competence. In a study by one of the authors (Pintó & El Boudamoussi, 2009) teachers' perception of the competencies demanded by PISA tasks were analysed, showing that, for Spanish teachers, only those tasks devoted to the competence "describing, interpreting and explaining scientific phenomena" were widely recognised. Approximately half of the teachers had greater difficulties in identifying the actual purpose of PISA tasks assessing sub-competences such as "understanding of scientific investigations" and particularly about "the use of scientific evidence". This latter sub-competence appeared least frequently identified as an assessed PISA task and least expected to pose difficulty for pupils in the science classroom.

The literature shows that students have difficulties with scientific inquiry skills, in particular regarding the collection and interpretation of evidence. For instance, analysis of the evidentiary competence in 6th graders revealed that students' understandings and reasoning skills of scientific evidence and the data collection process was quite weak in several respects (Jeong, Songer & Lee, 2006). However, the research of Pintó and El Bodamoussi shows that teachers are in fact not aware of these difficulties so it is reasonable that they do not include these contents in their teaching. Without working directly with teachers in this area, it is very difficult to see how pupils' results will improve. In this sense, we agree with the authors that implications of their study signal the importance of educating teachers on "scientific processes" and "the need to develop activities or identify existing ones that would help teachers put more emphasis on scientific processes, rather than on scientific knowledge and concepts themselves" (Pintó & El Boudamoussi, 2009, p. 39). This became the focus of GIMMS Spain as it linked the education of pre-service and in-service teachers and the elaboration of these activities within a mentoring framework, with teacher educators, for teaching and teacher education.

Work plan for GIMMS Spain

GIMMS Spain involved both in-service and pre-service teachers, working alongside teacher educators/researchers, in a particular problematic context – the teaching and learning of the competence "using scientific evidence". This was being done within the context of a very insufficient model of initial teacher education and training. In the following sections, we discuss this problematic context, elaborate on the content and describe the strategy used in GIMMS Spain for both improving teaching and teachers' education and designing adequate teaching and learning activities for teaching scientific competence "using scientific evidence". The work plan is finally presented as the series of actions developed to accomplish this purpose.

Context of GIMMS Spain: the end of an era of insufficient pre-service education and training for science teachers.

The perceived science teacher profile in the Spanish secondary school is that of a subject matter knowledge specialist with only a minor pedagogical education in the form of a pre-service course called CAP (Certificate of Pedagogical Aptitude). This CAP course is a post-graduate course, organised by teacher education institutes in universities, as a pedagogical and didactical course with a mentoring component (see description at "National summary sheets on education systems in Europe and ongoing reform", Eurydice, 2009). In the Universitat Autònoma de Barcelona, the CAP course for future science teachers was organised according to the following structure:
- 4 credits (aprox. 40h) of face-to-face teaching of didactics of science/education
- 4 credits (aprox. 40h) of face-to-face teaching of general pedagogy
- 24 credits (aprox. 240h) of practice in state secondary schools-practicum – with mentor-teachers, who are in-service secondary school teachers with many years of teaching and mentoring experience.

Since its origin from the early 1970's the CAP course has been widely criticised as providing insufficient pedagogical education and training for future teachers. By the end of GIMMS Spain the CAP course was officially replaced by a master course (according to the Bologna structure) of one year duration consisting of 60 European credits (Eurydice, 2009). The main educational innovation and change comes from the extended role given to mentoring in the school-based practicum experience.

The teacher education and mentoring context for GIMMS Spain is that of the CAP course. However, the fact that the traditional teacher education model was being officially replaced and different roles were being demanded influenced the project in a number of ways. For instance, the number of students enrolling in the CAP course during these three academic years became greater than ever, as more student teachers wanted to be qualified before it became a longer, more expensive and demanding course. This was problematic in two senses. First, more student teachers enrolled in the CAP course without necessarily wanting to become science teachers. Second, and due to the increase in the number of pre-service teachers, more mentor teachers than usual were required. On the other hand, the change in initial teacher education had a positive effect: mentoring was going to have a greater regional policy role and, as a result, mentor teachers were keen to become involved in GIMMS Spain.

Content of GIMMS Spain: the scientific competence "using scientific evidence"

In a recent reflection about scientific literacy, Osborne (2007) posed two questions which science education needs to face: a) what should be the outcomes of a science education *for all*, and b) what kinds of learning experiences are required to attain those goals. We considered the achievement of scientific competence the main outcome for science education. The second question, how to do this – and how to help teachers do it – is what drives our work with student teachers and their mentor teachers to promote teacher professional change and development in the science classroom.

Scientific competence has different definitions, all of them involving scientific knowledge but making clear the need to move beyond *having* knowledge towards *being able to use* knowledge (Perrenoud, 2008). A well-known framework for the definition of scientific competence has been used by the PISA 2006 study (Fensham, 2007).

For the purposes of PISA 2006, scientific literacy refers to an individual's:
- Scientific knowledge and use of that knowledge to identify questions, acquire new knowledge, explain scientific phenomena and draw evidence-based conclusions about science-related issues.
- Understanding of the characteristic features of science as a form of human knowledge and enquiry.
- Awareness of how science and technology shape our material, intellectual, and cultural environments.
- Willingness to engage in science-related issues and with the ideas of science, as a reflective citizen.

In this model, scientific literacy is characterised as consisting of four interrelated aspects: context, knowledge (*of* and *about* science), attitudes and competencies (Figure 1). Within the PISA 2006 framework scientific competence implies to be able to: identify scientific issues; explain phenomena scientifically; and use scientific evidence.

> "Scientific literacy is the capacity to use scientific knowledge, to identify questions and to draw evidence-based conclusions in order to understand and help make decisions about the natural world and the changes made to it through human activity". (OECD, 2004)

Aspects of the competence "use of scientific evidence" have been largely analysed and elaborated in the literature within different areas of research. The fields of research on practical work, epistemology or the nature of science, inquiry, argumentation or decision making, among others, have researched the role of evidence in science and society, in particular regarding how to obtain scientific evidence and how to produce arguments and conclusions based on this evidence (Osborne, Erduran & Simon, 2004). The PISA 2006 framework, "use of scientific evidence" refers to three concrete aspects:
- interpreting scientific evidence and communicating conclusions
- identifying the assumptions, evidence and reasoning behind conclusions
- reflecting on the societal implications of science and technological developments

and may involve "selecting from alternative conclusions in relation to evidence; giving reasons for or against a given conclusion in terms of the process by which the conclusion was derived from the data provided; and identifying the assumptions made in reaching a conclusion. Reflecting on the societal implications of scientific or technological developments is another aspect of this competency" (OECD, 2007, p. 30).

In GIMMS Spain we used the PISA definition for the competence "using scientific evidence". However, particular to the Spanish context was the focus on the knowledge of science and the knowledge about science. In the Pinto and El Boudamoussi study, teachers showed difficulty in differentiating tasks between learning mainly science and learning mainly about science. For this reason we focused with teachers on the competence "using scientific evidence" with attention to knowledge of science and about science.

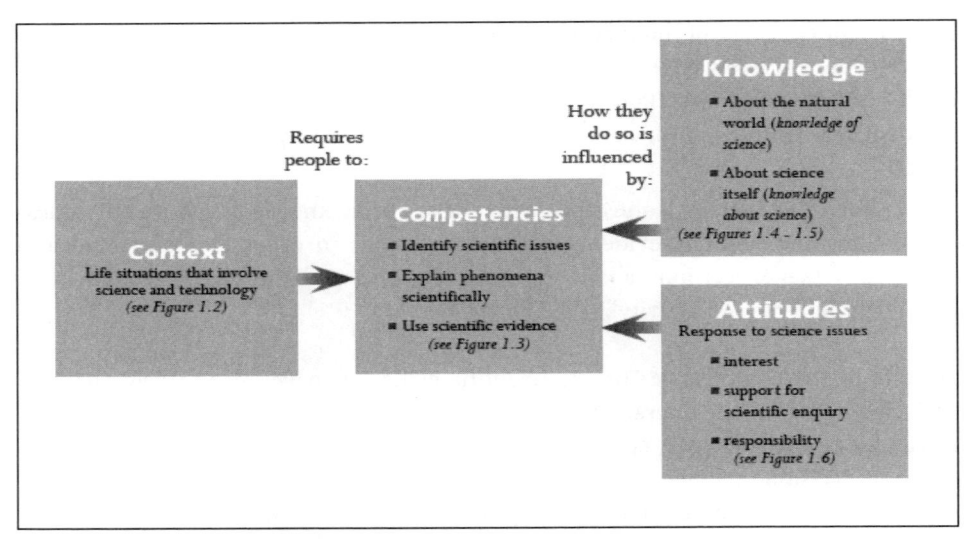

Figure 1:   Scientific literacy as characterised by PISA 2006.

Strategy of GIMMS Spain: goals, research questions and actions undertaken

GIMMS Spain had different goals regarding its development and research purposes. First, the goal of GIMMS Spain was to support in-service and pre-service teachers' development of teaching the scientific competence 'using scientific evidence' in their classrooms. A new mentoring framework, based on co-learning between mentor teachers, student teachers and teacher educators/researchers, was proposed.

Second, GIMMS Spain had a research-practice component, using the rich context of GIMMS activities as a particularly fruitful research scenario for illuminating future teaching and teacher education practice. In this sense, GIMMS Spain analysed the perspectives that both pre-service and in-service teachers had regarding the competence 'using scientific evidence'. For this second purpose, a research question was explored and results published at both national and international level (Couso & Pintó, 2009a, b). The research question was: what aspects of the competence 'using scientific evidence' do mentor and mentee teachers include in teaching and learning activities? To accomplish both goals in GIMMS Spain a series of actions were undertaken in the timeline of the project (see the following box).

2006–2007: Identification of a problematic area in Science Education in Spain
- Elaboration of the framework for the project
- Contact with the teacher and introduction to rationale
- Development of goals, research questions and work plan
- Definition of national project
2007–2008: First cycle for the new Innovation and Mentoring model
- Design and implementation of a series of *teacher education meetings* with the project group of mentor teachers.
- Design and critical reflection about *teaching and learning activities* for "using scientific evidence" by mentor teachers.
- Elaboration and updating of a *Moodle platform* for communication and exchange of materials with mentors.
- Design and implementation of project-based *pre-service training sessions* to mentee teachers on scientific competence
- Design and implementation of *teaching and learning activities* for "using scientific evidence" by student teachers and their mentors.
- *Pilot analysis* of in-service and pre-service teacher tasks (development of the analysis instrument)
2008–2009: Second cycle for the new Innovation and Mentoring model
- Continuation with *teacher education meetings* with teacher mentors. Critical reflection on year one student teachers' activities.
- Design and critical reflection about *teaching and learning activities* for "using scientific evidence" by mentor teachers.
- Updating of the *Moodle platform* for communication, support and exchange of materials/ideas with mentor teachers.
- Implementation of project-based *pre-service training sessions* to mentee teachers on teaching competence (new group of mentee teachers)
- Design and implementation of *teaching and learning activities* for "using scientific evidence" by student teachers and their mentors.
- *Analysis* of in-service and pre-service teacher tasks
- Design and implementation of a *teacher questionnaire* about mentoring model

Educational innovation and change and mentoring

Experienced in-service teachers in GIMMS Spain in Catalonia were, most of them, already mentor teachers of student teachers. This meant that, despite reform efforts, the mentors who would guide those student teachers' *practicum* would not likely use teaching that promoted the development of competences, and neither, according to our results, would fully understand this purpose in the teaching tasks designed by the student teachers. In this sense, for the particular curricular innovation being advanced in Catalonia via the new curriculum – the introduction of scientific competence – and within the particular problematic context evidenced by research – lack of both awareness and knowledge about the teaching and learning of the sub-competence "use of scientific evidence" by mentor teachers – it seemed a challenging strategy, for GIMMS Spain, to attempt to link innovation with the mentoring model within the existing initial teacher education programme.

However, in a traditional mentoring programme in which teacher mentors do not conceive themselves as co-learners and powerful novice-expert status relationships are the norm, the introduction of educational innovation and change as discourse, between teachers across their professional lifespan with teacher educators and others, becomes seriously affected, as mentor teachers do not traditionally integrate any new knowledge coming from mentee teachers. In a similar way, a view of mentor teachers as experts that know "all that is needed to know" to train new teachers means that change of practice is not required even when new knowledge is being introduced from science educators. In this way the mentor-teacher educator relationship gets reduced to the role of peer control and assessment of un-knowledgeable mentee teachers.

The introduction of educational innovation and change through a broad-based mentoring programme needs a new mentoring model that emphasises critical reflection and co-learning among all agents involved – mentee student teachers, experienced teacher mentors and university teacher educators/researchers. As other authors have beautifully expressed, there is a "need for synergistic and beneficial mentoring practices among different professionals" (Mullen & Lick, 1999, p. 4). In the field of particularly demanding innovations such as the teaching for the achievement of scientific competence, which is challenging for all education professionals – teacher educators/researchers, teachers and student teachers alike – this need is even stronger. As the same authors point out "if researchers, teachers, and other professionals are to develop, they need to work together to highlight partnership forms of new learning and reciprocity" (p. 15).

The GIMMS project in Spain, during the CAP course of 07/08 and 08/09, invited a group of in-service teachers of physics and chemistry that were "traditional mentors" to participate in an initiative of innovative school practice: the introduction of teaching for the achievement of the scientific competence "using scientific evidence" in their schools by using their role as mentors and exploring a new framework for mentoring. This new model used notions of *critical co-inquiry* supported by teacher educators/researchers (Wang & Odell, 2007) and, more importantly, it supported the concept of co-learning between mentors and mentees.

Cooperation and settings

The described rationale for educational innovation and change required particular cooperative settings to be established. A summary of the collaborative learning and partnership relationships established in GIMMS Spain is found in Figure 2. It is based on the agreed GIMMS team framework decided at the start of the project (Chapter 1). Double arrows represent reciprocal opportunities for collaboration and feedback in different settings:

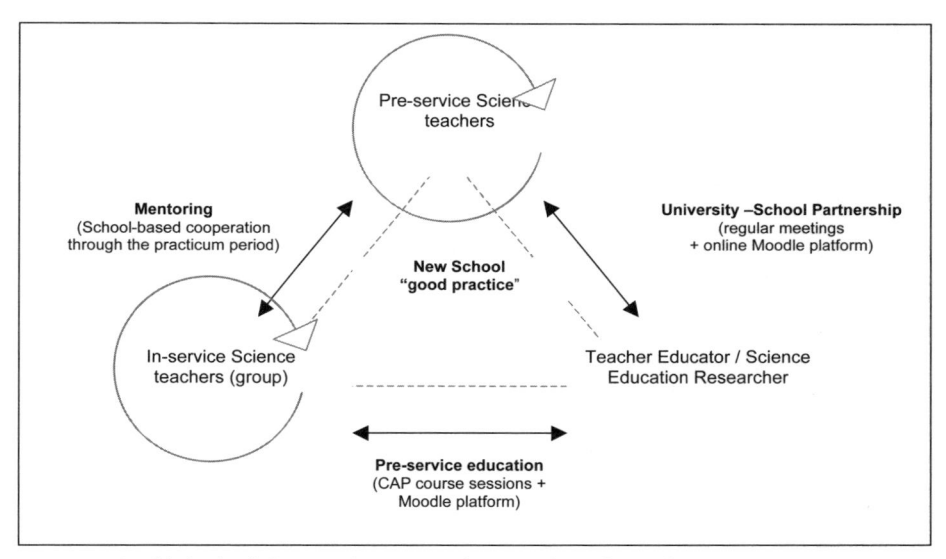

Figure 2: Established collaborative learning and partnership relationships

- face-to-face teacher educator – mentor teacher meetings
- face-to-face teacher educator – student teacher seminars
- school-based collaboration between mentor and mentee teachers
- on-line collaboration via a Moodle platform among all agents.

The settings, despite their richness, could have been improved. It would have been very beneficial to organise some reflective seminars were all participants (student teachers, teacher educators and mentor teachers) could meet, discuss and inquire together on aspects of the project. However, the contextual constraints of the project did not allow such meetings to take place, as there was a need for the actions undertaken to be compatible with the existing teacher education programme in which student teachers were officially enrolled. The shortness and rigid schedule of that course constrained the possibilities of global three-way cooperation. In the case of the mentor teachers' group, there were also minor changes in the composition of teachers and mentors from group to group and year to year. In GIMMS Spain there were over 20 mentor teachers, approximately 50 student teachers and three teacher educator/researchers involved. Table 1 summarises the actual number and type of participants in the project.

Table 1: Number of type of GIMMS Spain participants.

2006–2008, year 1 and 2
Data from semesters 1 and 2
22 physics and chemistry in-service teachers (confirmation of participation in the reflection meetings from 11) from 21 schools. 47 physics and chemistry student teachers attending the CAP course 2 science teacher educator/researchers 1 science teacher educator/researcher(GIMMS Spain National Coordinator)
2008–2009, year 3
Data from semesters 1 and 2
34 physics and chemistry in-service teachers (confirmation of participation in the reflection meetings from 11) from 21 schools. 55 physics and chemistry student teachers attending the CAP course 2 science teacher educator/researchers 1 science teacher educator/researcher

Gender awareness

Gender sensitivity was not a particular focus of GIMMS Spain. First, teachers were grappling with new competency-based curriculum requirements in Catalonia and did not feel that they could include an additional focus at this time. One innovation at a time seemed more than enough. Besides mentor teachers, both male and female, were reluctant to reflect, inquire and design materials specifically for gender awareness.

Second, PISA results for Spain had indicated there were little or no gender difficulties in some aspects of the education system. For example, the OECD report "Equally prepared for life" (OECD, 2009a) where PISA results on gender are summarised, it stated that there were no gender differences among girls and boys for primary education in Spain. Regarding secondary education, PISA results on boys and girls of 15 years-old situated Spain among those countries where there were only moderate gender differences in two measures, regarding performance or attitudes to science.

Third, the teaching and learning activities designed in GIMMS Spain were highly contextualised and included a strong linguistic (reading, writing) and reasoning component (argumentation, elaboration of conclusions). These aspects have been argued as necessary to include girls in the science classroom. In particular, the PISA 2006 competency framework for the teaching of science education, used in GIMMS Spain, stresses the idea of "teaching for understanding" and connecting knowledge instead of learning isolated facts, which research has highlighted as a very significant factor for girls favouring science (Zohar, 2003).

Data collection and analysis

In GIMMS Spain data was collected from mentor teachers and mentee teachers in two separate cohorts over the lifetime of the project.

Process of data gathering from the first cohort of mentor teachers

Mentor teachers, in the first cohort, found implementing the competency a challenge and began to slowly perceive the potential of a different type of mentoring relationship with student teachers. Their comments are summarised in the following box.

> Mentor teachers find the idea very interesting to introduce the work on the scientific competency "using scientific evidence".
>
> They find it a challenge and openly comment that this is something "they never do in their science teaching".
>
> They find a challenge to help pre-service teachers in doing this, when they are not very much confident in how to introduce this innovation in practice themselves (they position themselves as experts in the mentoring relationship).
>
> They also start to grasp the potentiality of the proposed new mentoring model by realising that the student teachers, having been able to work on this longer and with the researchers support in the CAP course, will be also a source of input and help for them.
>
> For introducing this new practice, they recognised the importance of having different examples and suggest they share their materials in an internet platform.

Process of data gathering from the second cohort of mentor and student teachers

In year two of the project a group of n=22 mentor teachers were invited to individually design activities for teaching the competence "using scientific evidence". They received feedback from teacher educators/ researchers and colleagues via a Moodle platform. However, few activities were designed and mentors' informal comments showed their feeling of a lack of experience. Student teachers (n=42) designed similar teaching and learning activities, many of them implementing these activities in their school-based setting.

Process of data gathering on the third year of the project

The reflection and analysis activities and tasks produced in year two served as scaffolding for mentor teachers in the third, and final year of GIMMS Spain. These were critically discussed and reflected upon with teacher educators/researchers. By discussing examples of practice which they helped to design, teachers' critically reflected on their practice and gained experience in the field designing exemplary activities themselves. 13 mentor teachers designed and exchanged activities with colleagues, and the teacher educator, via the Moodle platform. The new group of year three student teachers (n=55) were also required to complete the same task of designing a teaching and learning activity. The activities/tasks were designed in collaborative groupings of 2 to 3 student teachers. These were again implemented in their school-based practicum with mentor teachers. A total of 19 activities/tasks were produced.

Research instruments

Instrument to analyse the level of competence

To analyse the level of scientific competence demanded from the teaching and learning activities an analysis instrument was designed. Following the project rationale, this instrument measured the level of contextualisation of the activity, the level of competence demanded in the work with evidences and the level of competence demanded in the work with arguments and conclusions from evidence. Figure 3 shows a summary of these various levels – contextualisation, argumentation, evaluation – for analysing mentor teachers and mentee teachers' activities/tasks.

Contextualisation

The type and level of contextualisation of the designed teaching and learning activities was analysed according to the PISA framework for characterising the context of activities. The research instrument used characterised the context of the teaching and learning activities regarding different aspects – ethical, political, technological – and dimensions – personal, social, global – addressed. The instrument also measured if contextualisation was only done at the beginning of the activity, as a kind of motivational input, or if it was continued throughout the activity/task.

Evidentiary and argumentative competence

The levels of competence demanded by the teaching and learning activities followed the classification proposed by Wilson & Chalmers (1988). The authors classified four distinct levels of tasks/activities: literal tasks, inferential tasks, creative tasks and evaluative tasks. For literal tasks the student teachers have only to report or describe directly what has been given to him or her. With inferential tasks student teachers have to infer from different sources of knowledge to develop the tasks. Creative tasks go further and demand from the student teacher the generation of a creative or new type of knowledge. Finally, evaluative tasks are those in which student teachers have to judge and develop their critical view. Based on this classification system, we elaborated a framework for analysing levels of competence in the use of evidence and conclusions demanded in the teaching and learning activities/tasks. The final version of this framework is summarised in Figure 3.

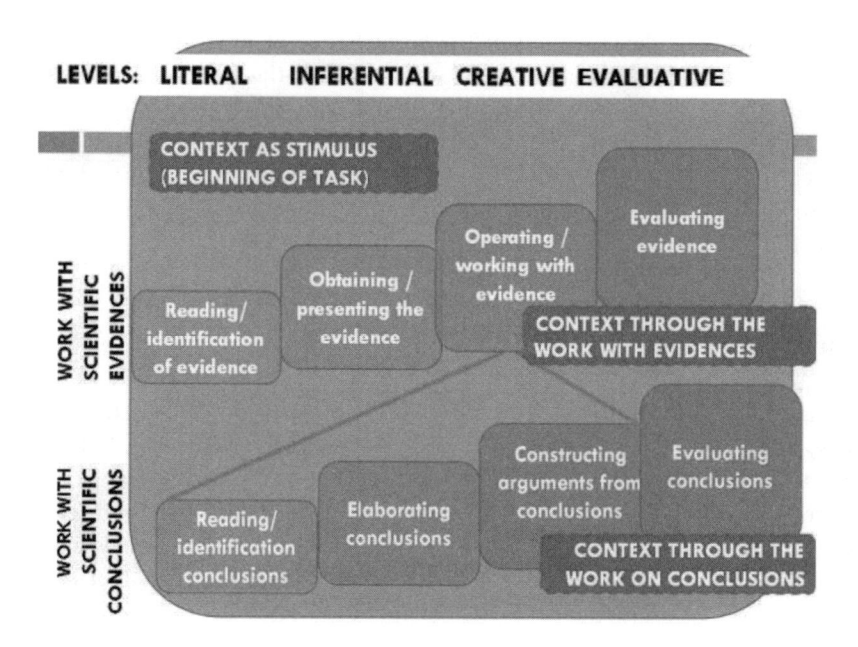

LEVELS: LITERAL   INFERENTIAL   CREATIVE   EVALUATIVE

Figure 3:   GIMMS Spain instrument for analysis of levels of scientific competence

The details of the types of tasks included in each of the levels were based on the literature work on concepts of evidence of Gott, Duggan & Roberts (without year) and also included aspects of the nature of science. To classify particular tasks ideas from the literature on argumentation, in particular from those studies using the Toulmin argument pattern, were used. Using this framework, a detailed instrument to characterise mentor teacher and student teacher designed teaching and learning activities was constructed and published elsewhere[3]. The purpose of this analysis was to evaluate to what extent, using the new mentoring model linked with innovation, both student teachers and mentors have improved their understanding and use of the competence "using scientific evidence" in their teaching.

Mentor teachers' questionnaire

A questionnaire for mentor teachers was designed and delivered by the end of the project. It asked for aspects of their understanding of the competence "using scientific evidence", their teaching of this competence in class, and the introduction of this competence to their student teachers during the practicum. Finally, they were asked about their perceptions of how the different interactions with the various communities – with teacher educator/researchers, mentor teachers and men-

---

3   Master thesis work titled "Análisis de actividades de enseñanza-aprendizaje, propuestas por futuros profesores de ciencias en formación inicial, para el desarrollo de la competencia "utilización de pruebas científicas" en el aula de ciencias" done by Josvell Saint-Clair, under the supervision of the author D. Couso. 2009, Universitat Autònoma de Barcelona.

tees – helped in their development of the competence of teaching "using scientific evidence". Teachers' questionnaires were analysed with the intention to evaluate to what extent, by using the new mentoring linked with innovation model, mentors improved their understanding and use of the competence "using scientific evidence" in their teaching and how, according to them, the different interactions within each of the *communities* involved fostered and supported their learning and development regarding the mastering of the competence in the classroom.

Change in practices

The change in practice stimulated by GIMMS Spain may be discussed using both a process and product orientation. Regarding the process orientation, both mentor teachers and the science educator/researcher reported as a positive experience their mutual collaboration. These positive outcomes are evidenced in the comments from teachers both in the first meeting (Table 1), the Moodle platform, emails and interactions in face to face meeting; and in the teachers' questionnaires. In fact, teachers were keen on a continuation of this collaboration after the GIMMS 2006–2009 lifespan. As a result, a national project was submitted in 2009 (COMPEC; Ref. EDU2009–08885) and it received substantial national funding for the continuation of work initiated in GIMMS Spain 2006–2009.

How mentor teachers have developed regarding the teaching and mentoring of the competence "using scientific evidence"?

GIMMS Spain introduced innovative teaching in the field of scientific competencies to both mentor and mentee teachers by involving them in designing curriculum materials – teaching and learning activities – to work on the competence "use of scientific evidences" in their physics and chemistry classrooms. Taking into account the Spanish context described this was not an easy task neither for teachers – who have shown to have difficulties regarding this particular competence – nor for their pupils – who performed under the OECD mean according to PISA results in the physical sciences.

GIMMS Spain does an interesting change in practice. The change was tracked from no recognition of the competence (Pintó & El Boudamoussi 2009 results) to acknowledgement with poor implementation to the actual design of tasks and activities that support the teaching of the competence:
- acknowledgement of the competence without being able to design materials to introduce it to their students (year two of the project)
- actual design, implementation and re-design of teaching and learning activities plus mentoring of student teachers on this issue (year three of the project).

Regarding the product orientation of GIMMS Spain the activities themselves are a necessary resource outcome of the project, as the competency framework explored is mandatory – due to the new curricular reform – and teachers in Catalonia lack sufficient examples of these materials. A range of these activities were developed in GIMMS Spain by mentor teachers and student teachers. Figure 4 shows an example of one of these activities.

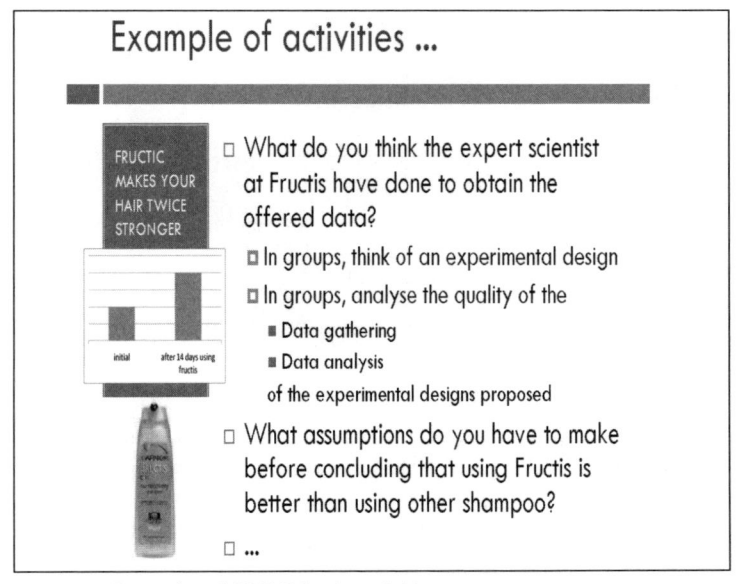

Figure 4: Examples of GIMMS Spain activities

What mentee teachers learned about teaching "using scientific evidence"?

We analysed the mentee teachers' activities in terms of the level of scientific competence summarised already in Figure 3. This allowed us to conclude that GIMMS Spain was a good learning and mentoring experience for the student teachers' participating.

Results regarding levels of contextualisation

Most of the activities designed by student teachers to introduce their pupils to "using scientific evidence" were contextualised at the highest level, as shown in Figure 5 and 6. This implies that, regarding contextualisation, the activities designed by mentee teachers:
- maintained context until the end of the activity, from the introduction of the activity to the work with evidence to the elaboration of conclusions.
- chose mostly between scientific and everyday scenarios

- included other dimensions in addition to the scientific one, such as the socio-political, economical, ethical and/or technological dimensions, both at the personal and social level, in their contexts.

These were very positive results and reflected the effort made both in the teacher education seminars with mentee teachers and in the seminars with mentor teachers, regarding the importance of contextualisation in the competency framework.

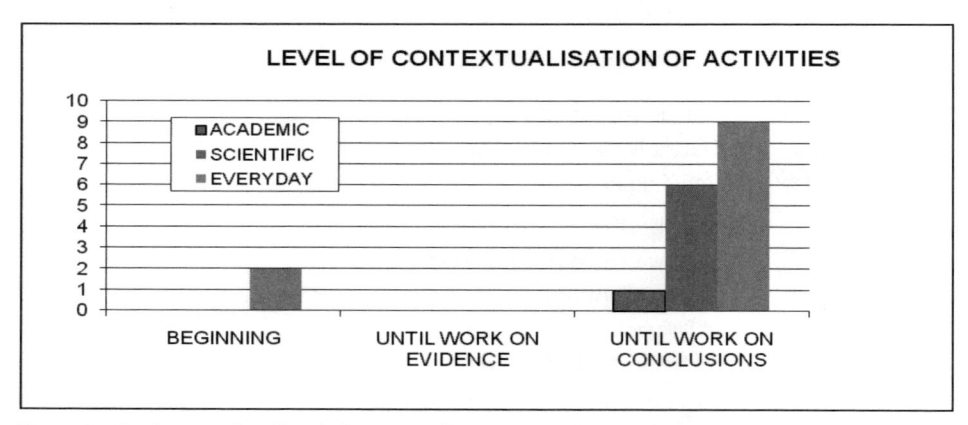

Figure 5: Student teachers' level of contextualisation in GIMMS Spain activities

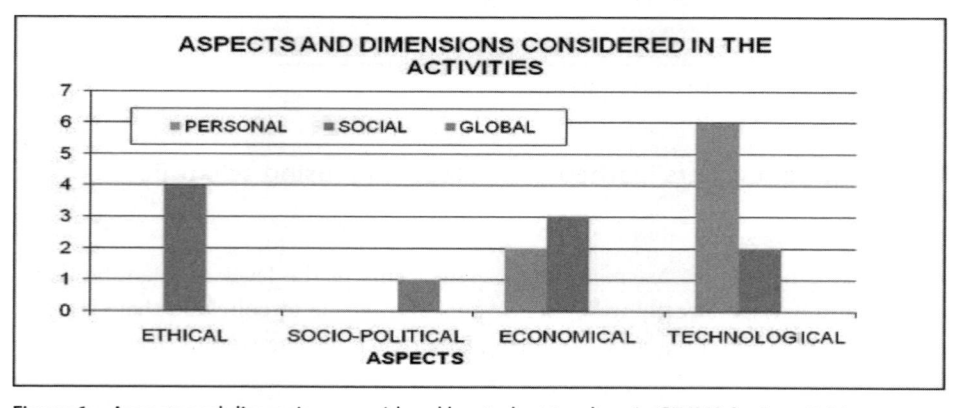

Figure 6: Aspects and dimensions considered by student teachers in GIMMS Spain activities

Results regarding work with scientific evidence

Regarding the work with scientific evidence, or the work in evidentiary competence, results were somewhere in the middle ground, as shown in Figure 7.

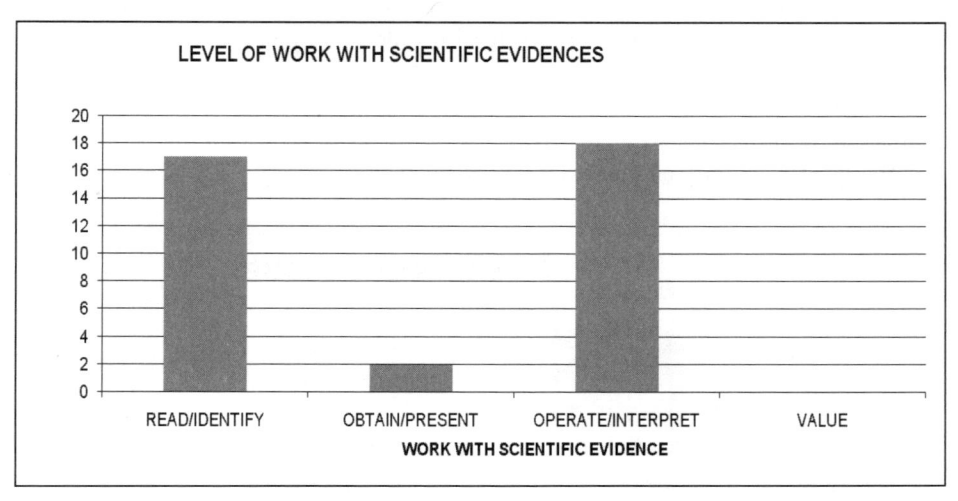

Figure 7: Student teachers' level of work with scientific evidence in GIMMS Spain activities

However, student teachers' activities were found to be mostly low competency activities at the literal level, asking pupils only to identify evidences (mostly in text or tables) or middle level, asking students to operate with evidences. Interestingly, there was an almost non-existence of tasks in which students had to obtain evidences themselves, either in practical work or by searching the internet or other sources. The operation most often included was that of answering questions from data, questions which were mostly inferential (see Figure 8). Overall, it showed a medium level demand in the student teachers' activities, which would need to be improved if more creative or evaluative tasks were to be posed. We see here a possible problematic issue that would need further research and exploration.

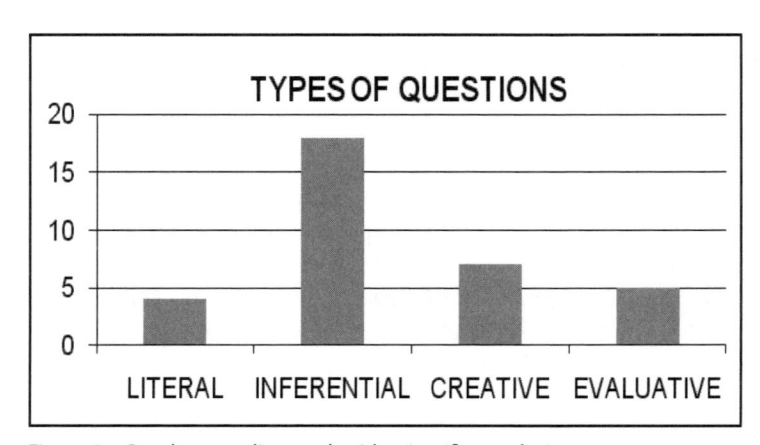

Figure 8: Results regarding work with scientific conclusions

According to our GIMMS Spain framework for analysis, presented in Figure 3, the higher level of scientific competence in "using scientific evidence" occurs when this evidence is used to obtain conclusions and develop arguments. In this sense, the analysis of student teachers' activities regarding the level of work on the argumentation aspect shows (see Figure 9) a low level profile of competence. In the activities designed by student teachers, pupils were merely asked to select the proper conclusions from a group of given ones or to elaborate on their own conclusions, but never to create new knowledge with their conclusions by communicating them, or to evaluate the quality, validity or reliability, of conclusions.

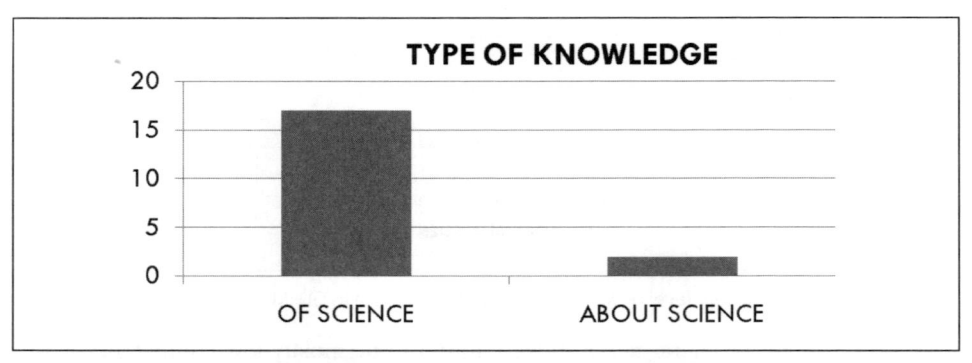

Figure 9: Use of types of knowledge made by student teachers' in GIMMS Spain

Analysis showed that, regarding the level of argumentation demanded, it was all at level one. Arguments were constructed directly from data but without demanding backing, for example, a theoretical justification or a rebuttal argument (against the main argument). In this sense, results seem to suggest the need to include specific education and training on argumentation for both experienced mentor science teachers and mentee student teachers, in order to improve the level of scientific competence demand from pupils.

Results regarding the evaluative dimension

An important outcome of the analysis of student teachers' teaching and learning activities is the lack of the evaluative dimension, which is the higher level in scientific competence to be achieved. Quality of data, of experimentation, of the conclusions drawn etc. was not addressed, which has important implications regarding pupils learning of the nature of science and the competence "using scientific evidence". In fact, when analysing the type of knowledge activities demand (Figure 10), most dealt with knowledge of science as distinct from knowledge about science. It appears that knowledge about science is either neglected or expected to be learnt elsewhere.

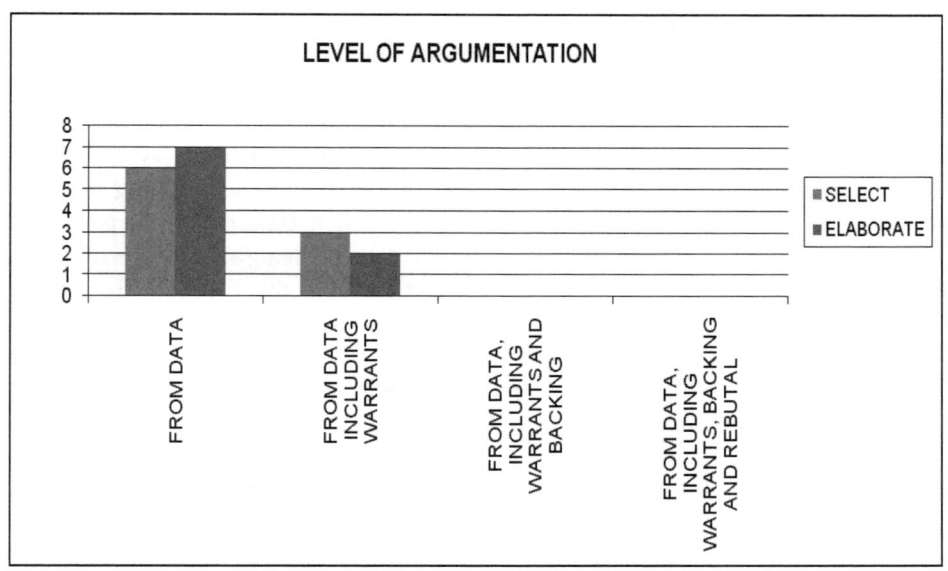

Figure 10: The various levels of argumentation used in GIMMS Spain

Discussion, results and some implications for future research

The results from GIMMS Spain show improvements regarding teachers' understanding on the competence "using scientific evidence" when compared with previous results in our context (Pinto & El Boudamoussi, 2009). The tasks designed also showed better quality than those from previous courses in which the new mentoring model was not used. The overall results may be stated as follows:
- Mentor teachers participating in GIMMS Spain evolved from a state of no recognition of teaching and learning activities to developing the scientific competence "using scientific evidence" to becoming able to design these teaching and learning activities themselves and to support mentee teachers in the process of designing and implementing these activities in the classroom.
- These teaching and learning activities, when carefully analysed, showed that mentee teachers' were able to adequately contextualise activities for the development of scientific competence. These sorts of activities seemed to foster, in fact, a degree of high quality contextualisation.
- Either due to lack of knowledge on how to promote the competence or by considering these activities too difficult for their pupils, mentee teachers designed activities of low or medium levels of difficulty for the competence 'using scientific evidence'.
- The critical view was missing in the activities designed by both mentor teachers and mentee teachers which can have serious consequences. This implies a need for a particular focus on this issue in future innovation and mentoring projects.

The innovation topic, in agreement with the new curricular reform, was widely accepted by GIMMS Spain mentor teachers in Catalonia and improvements were noted in their practice regarding the competence framework. However, the same did not occur regarding their recognition of the new co-learning relationship advocated with their mentees. Very few mentor teachers recognised at any deep level their role in the development of awareness or sharing of experience in designing teaching activities for teaching "using scientific evidence" in their classrooms. Despite some experienced teachers' recognition of the importance of being mentors to improve their practice of teaching, they generally did not associate this fact with the mentee teachers' contribution or any element of co-learning with the student teachers. In GIMMS Spain the experienced mentor teachers were more open to the co-learning opportunities found within the open 'public spaces' for discussion with other mentor teachers and the teacher educator/researchers.

These constraints and difficulties regarding the mentoring model in GIMMS Spain are not entirely surprising. For the authors of this chapter despite many years working within teacher education, the link between educational innovation, change and mentoring that GIMMS 2006–2009 framework conceptualised was new and innovative. Collaboration with other European partners helped to identify the problems particular to our own context – in the midst of a curricular reform – and also to identify common shared values and issues.

The international literature on science education shows that the conception of school-based mentoring among science teachers, both from the mentor and the mentee perspective, is often closer to an "apprenticeship model" with expert and novice. Personal support and co-learning are harder to be found. A co-learning mentoring framework, as advocated by GIMMS, is grounded in the belief that the knowledge, experiences and perspectives of all actors involved – teacher educator/researchers, mentors and mentees – offer crucially important insights for deeper understanding of professional practice, values and pedagogical innovation. This remains the central challenge for educational innovation and change in teaching and teacher continuing education.

Chapter 8:
Czech Republic: constructivist approaches to innovation in one school-university partnership

Eva Volná, Hashim Habiballa, and Rostislav Fojtík
University of Ostrava, Czech Republic

Traditional learning and teaching in the sciences and mathematics

In PISA 2006 (OECD, 2007) the Czech Republic performed well in both the science and mathematics domains. Concerning science literacy among GIMMS project countries the Czech Republic was above the OECD average and second only to Germany in terms of student performance (Table 1, science). Concerning mathematical literacy among GIMMS project countries the Czech Republic was also above the OECD average and second only to Denmark in terms of student performance (Table 1, math).

Table 1: PISA 2006 science and mathematics results across the GIMMS countries*.

GIMMS country	Mean Scores Pisa 2006	
	science	math
Czech Republic	513	510
Denmark	496	513*
Austria	511	505*
Germany	516	504*
Ireland	508	501*
Spain	488	480*
OECD average	496	496*

* Gender difference in mean scores is statistically significant with 95% likelihood

In the Czech Republic, at lower secondary education (11-15 year olds), there is a traditionally inherited prevalence of transmission type teaching and schooling. Teachers tend to rely heavily on verbal forms of teaching with encyclopaedic types of knowledge prioritised. To a lesser extent, teachers implement activities aimed at the practical study of facts, teaching through individual perception and application of individual experience. Orientation towards pupils' performance and test score achievement is prevalent, with insufficient differentiation for individual personal abilities and pupil-centred teaching.

Possibly these results may have been due to a strict organization of lesson planning and classroom organisation and specific use of technologies. Teaching of science generally shows more effective use of specialised classrooms, such as laboratories. However, laboratory practice is not scheduled as a regular activity. Pupils are given lots of subject matter knowledge but are often uncertain as to its application in practice and its relationship to everyday life and living. In the mathematics lessons, teaching aids and technology are oftentimes used but to a limited extent. Pupil-centred teaching and differentiation in assigning tasks is not strongly demonstrated. There is an increased interest nationally in pupils using new information technologies. In this regard, upgrading of information technology (including software) in accordance with its dynamic developments in practice has become increasingly an issue in the classroom and school setting.

Problems are occurring in ensuring science and mathematics teachers, and the teaching staff generally, have appropriate pedagogical knowledge and, even more importantly, professional qualifications. The issue of teacher qualification is specifically a problem at isolated village schools in peripheral locations with a limited transport service. The equipment facilities enable, in some cases, full performance of the curriculum, however, there are serious difficulties in other cases. As facilities are variable the majority of schools are not significantly upgraded and do not have adequate teaching aids or up-to-date technology. This gives the background and context for GIMMS Czech Republic. The project operated as an exploratory school-university partnership seeking to develop educational innovation and change in the ways teachers worked more interactively with pupils in their classrooms.

New innovation trends

In 2009 research was conducted in secondary schools. The sample size for this research study was 270 teachers with different teaching qualifications. The study was conducted at grammar schools, secondary vocational schools and vocational schools. One of the key questions was the number of students per computer and the use of different technologies. It was found that ten students, on average, share a computer. However, depending on the type of schools, facilities varied. The best situation was found to be at secondary vocational schools, where less than eight students share one computer. There were about twelve students per computer at grammar schools and twenty students per computer at vocational schools.

Similar results were also found in the use of technologies, such as data projectors and interactive whiteboards. It appeared that school facilities were mostly adequately resourced, having at least their minimal required number of technologies. However, the application of these technologies showed an entirely different situation. The inquiry showed that even at the relatively well-equipped schools there were teachers, who for example, never used interactive whiteboards or data projectors in their lessons. On the other hand, computer science teachers were found to use the modern computer equipment in almost every lesson. Some teachers, especially mathematics teachers and science teachers, endeavored to use new methods and information and communication technologies, but in their case it was found that computers were mainly used as a presentation tool. This latter finding is in keeping with earlier claims of the prevalence of transmission models of teaching, learning and assessment. Overall there appears to be a growing number of teachers who are as a minimum requirement use the computer, and a variety of technologies, in preparation for their teaching. However, computers were less applied in the following scenarios:
- Testing and examinations.
- Administrative solutions such as the running of schools and school systems.
- Communication networks with parents and students.
- Artistic production and in school laboratories.

These findings have implications for the development of innovation in GIMMS and also, have policy implications, for the continuing education of teachers, which clearly needs to focus not only on the mastery of working with modern technologies, but mainly on their use in developing progressive pupil-centred teaching methods and approaches.

Czech GIMMS research and development

The goal of Czech GIMMS was to develop innovative gender-proofed practices in the teaching of science and mathematics using constructivist approaches (Driver & Oldham, 1986; Driver & Bell, 1986; Matthews, 2000) of learning in one school-university partnership. Czech teachers cooperated with the project teacher educator and researchers in the areas of mathematics, physics and computer science, at lower secondary school. GIMMS was established as an exploratory study of the 'new' and innovative roles of the teachers and the resulting changes in motivation and/or practices.

In GIMMS Czech Republic there were four co-operative schools and eight teachers, across three subject areas – mathematics, physics and computer science. The teachers sought to develop constructivism in their approaches to science and mathematics teaching. Cooperative pupils were between the ages of 10 and 16 years old. Over 250 pupils took part in the project. The following table shows the number of participating pupils by subject and by gender. In this chapter we consider the topics chosen by these teachers and the resulting approaches to teaching and learning adopted by these teachers, in dialogue with the teacher educator/researchers. We will also consider the findings and their implications going forward.

Table 1:   Participating pupils in GIMMS Czech Republic by subject and by gender.

Physics		Computer Science		Mathematics	
boys	girls	boys	girls	boys	girls
18	12	46	52	64	59

GIMMS Czech Republic: teaching in Physics

The topic of the physics lesson, used in GIMMS Czech Republic, was "quantities and their measurement". This topic was not just passively explained, but pupils actively cooperated during all lessons, preparing classroom materials in advance and taking part in a number of experiential activities. Pupils tested quantity measurements in practice, for example, quantities of weights, length, and time. It was observed that boys were found to be very active during whole lessons and were interested in the new approach and motivated by the new topic. The teacher was motivated by the new approach and the active pupil preparation for the lessons. The teacher reported that pupils were found, using this co-planning approach, to memorize better and learning outcomes were generally found to be much better than with previous transmission approaches.

GIMMS Czech Republic: teaching in Mathematics

The topics of mathematics lessons used were "square and square number", "highest common factor and least common multiple", "geometry of conic sections", and "identical and similar mappings in constructive tasks". The topic "square and square number" was not found to be a difficult topic and pupils were ready for individual work and an active approach to exploring a square number definition. This was done using both individual work and work in groups, in an effort to offer both whole class teaching and individual teaching. Conclusions made by pupils were then presented in a public way to their teacher, and peers.

The teacher showed a number of different ways of looking for the highest common factor and least common multiple. This topic was explained using word tasks. The teacher made conclusions in active cooperation with the pupils. Pupils were required to learn the new subject matter by themselves and to take responsibility for their own learning. This new aspect of the curriculum was then practiced in groups and finally pupils solved the test on this topic by completing a computer assignment.

The topics "geometry of conic sections" and "identical and similar mappings in constructive tasks" were taught using *Cabri Geometry Software*. This software is similar to computer geometry building blocks – a modular system – which pupils operate by themselves. The approach integrates ICT into mathematics teaching and demands more effective work from pupils who have to solve tasks by themselves. Teachers reported that this teaching was found to be more attractive for pupils than traditional teaching approaches using rulers and drawing-compasses.

GIMMS Czech Republic: teaching in Computer Science

Computer science is one of the most popular subjects in the school curricula. The topics for computer science lessons used were "MS Word", "Text Editors", and "Project Works". Teachers observed and reported that both boys and girls were very active during all lessons. Boys appeared to be especially interested in the technical background and the girls appeared to prefer communication work such as interviews and work with information. It was observed that girls were better in theoretical learning and boys were better in practical computer application. The lessons with the topic "project works" integrated the computer science subject to project teaching. This involved crossing subject boundaries and making cross-curricular links between a range of subjects including computer science, creative education, Czech language, and English.

Pupils made leaflets for a school theatre performance, which were publicly presented. There was a busy working atmosphere during whole lessons and because of this group work lessons were noisier than usual. Working in groups stimulated discussion, pupils reactions were more spontaneous, and individuals, who previously were quieter and less involved, now joined the discussion. Teachers, involved in this aspect planned to continue these active ways of teaching into the future.

In conclusion, *sciences*, *mathematics*, and partly also *computer science* do not belong to popular subjects at secondary schools and primary schools. Using a variety of the pedagogical approaches, teachers taking part in GIMMS reported that the project helped to increase pupil's interest, understanding and motivation in these subjects. This suggests that exploratory school-university partnerships in the Czech GIMMS project have the capacity to improve motivation and understanding through the application of a variety of pupil-centred teaching methods and work practices.

Educational innovation and change in the Czech GIMMS project

The success of educational innovation and change in the Czech GIMMS project was measured by the improvement of learning outcomes and results from a comparative study of pupils' interest and motivation in the sciences and mathematics before the start of the project and after the project was completed. The project analyzed the processes in three subjects: physics, computer science, and mathematics. The key focus of the investigation was on the development of gender sensitive resources and materials and practices based on constructivist approaches, and reflection on the classroom experience using these materials.

In the project new and innovative approaches to teaching science and mathematics were realised. Teachers implemented the activities aimed at the practical study of facts but also moved teaching beyond this to teaching through pupil-centred perception and application of individual experience. Both a product and a process perspective was brought to bear on the project. The product perspective explored practical items such as the choice of topics and subject matter knowledge chosen by the teachers as their areas for exploration. The process perspective considered the teaching rationale and methodology for a pupil-centred and inclusive approach to teaching science, mathematics and computer science.

Data collection and analysis

Data collection in the Czech GIMMS project included teachers' teaching materials, field inquiry observations and questionnaires. Cooperation in the project was across a number of different settings and included a school-university partnership approach, a teacher-teacher partnership and a teacher-pupil partnership. The analysis was based on the apprenticeship model of mentoring by Maynard & Furlong (1995) and the extended aspect of this framework as agreed by the GIMMS team.

Educational innovation and change from the product perspective

Czech teachers suggested the topics that they taught using a range of constructivist approaches in physics (*quantities and their measurement*); mathematics (*square and square number, highest common factor and least common multiple, geometry of conic sections, and identical and similar mapping in constructive tasks*); and computer science (*MS word, text editors and project works*). This resulted in the design and development of a number of different classroom resources which were disseminated to other experienced teachers and student teachers.

Educational innovation and change from the process perspective

The process perspective included all efforts at moving pedagogical practices, from older transmission models of teaching, learning and assessment to more progressive pupil-centred models. Arguments were exchanged, in this school-university partnership, between traditional inherited practices of transmission teaching and the new and different type of teaching approach that is needed to work with pupils' lay theories and facilitate their emerging understanding. Constructivist pedagogical practices concerns itself with how to enable a process of constructing new knowledge from existing knowledge and pupil's lay theories. In this latter approach, and according to the theory of constructivism, a teacher cannot ignore the pupil's existing knowledge base and teachers need to begin with this, using a facilitation approach. The teacher's role is to build with their pupils consistent theories that are in agreement with the currently accepted scientific theory.

Using a constructivist approach teachers need to be able to question the pupil in order to find out what lay theory pupils' are currently using to understand the scientific phenomenon under discussion. They then attempt to guide the pupil to the 'correct' scientific theory.

A number of key principles with regard to teaching using the pedagogy of constructivism came to the fore during the Czech GIMMS research and development project, including:
- The teacher needs to articulate, regardless of teaching technique (lectures, labs, assignments) the cognitive change that she/he wishes to bring about in the pupils and structure the activity to achieve this aim. Merely transferring knowledge is not a meaningful aim.
- The teacher needs to delve underneath her/his own expert knowledge to expose the prior knowledge needed to construct a viable model of the material that he/she will be teaching. She/he needs to ensure that that the pupils have this prior knowledge.
- In any particular course you will be teaching a specific level of abstraction the teacher will need to explicitly present a viable model one level beneath the one that she/he is teaching.
- The teacher needs to assume, when a pupil makes a mistake or otherwise displays a lack of understanding, that the pupil has a more-or-less consistent, but non-viable, mental model. The teacher's task is to elicit this model fully and guide the pupil in its modification.

- The teacher needs to provide as much opportunity as possible for individual reflection (for example, analysis of errors) and social interaction (for example, group discussion).

Clearly, each educator/teacher needs to decide how to apply these principles in a concrete situation and context.

Interdisciplinary concept between mathematics and computer science

Currently an interdisciplinary approach seems to be the preferred option to improve computer science education. Habiballa, Volná & Fojtík (2006) reflected this interdisciplinary conception of computer science education, for example, between mathematics and computer science. A questionnaire was sent, through a web-page, to 142 schools from which 56 responses were obtained. The research findings bring an interesting observation: interdisciplinary teaching, across or within subject borders, appears to facilitate deeper learning than the mono-subjective approach.

We were able in this study to differentiate interdisciplinary teaching from two different points of view. The first perspective involves *interdisciplinarity in a broader sense*. From this perspective we might consider whole aspects where one subject interacts with another. For example the usage of ICT tools, mathematical software, learning-management systems, in mathematical education or the usage of mathematical principles in particular problem algorithms.

The second point of view we call *interdisciplinarity in a narrow sense*. Such an approach relates to specific issues treated in parallel. In our research we argued for teaching which is based on interdisciplinarity in this narrow sense. Our findings concluded that the number of teachers using this interdisciplinarity principle, based on the limited extent of our research, was low. Higher attention needs to be given to this issue. As it involves crossing borders, and meeting different subjects at the edge between them, it carries some of the necessary conditions for innovation and change.

Changes in practice in Czech GIMMS

During the Czech GIMMS project teachers began to change their practices from more transmission types of teaching to more pupil-centred activity based learning experiences. New topics were not only passively explained but involved pupils actively and cooperatively engaged during all lessons. Despite the fact that GIMMS pedagogical experiments were made in less popular subjects, such as mathematics, pupils were enthusiastic about their learning. Teachers reported that best results were realized in lessons, where information technologies were actively used. For example, this included pupil-learning in small groups, work with interactive whiteboards and testing using computers. Use of the interactive whiteboard as a shared workspace turned out to be important, not just as a presentation tool. Teachers described these lessons as busier, but with a better thinking and intellectual processing atmosphere. For example, solving mathematical tasks in groups was generally found to be easier and more motivating for pupils.

Pupils could also calculate and verify their work and presumptions with their classmates. This element of peer-to-peer learning is recognised as important for formative assessment. Pupils who normally did not contribute in class became involved in debate, because they felt stronger in the group. Mixed groups of boys and girls were generally found to be more active, girls were more active in evaluation of results while boys appeared better in calculations. Teacher observations indicated that pupils evaluated the lessons carried out by using information technologies more positively than less traditional forms of teaching. Pupils enjoyed these lessons and regarded them as a very pleasant change. For example, examination in mathematics through the self-directed computer test had a great response, 97% of pupils evaluated this format positively.

Changes in practice were found in a number of curriculum implementation areas, for example, for lesson planning, use of technology, use of activity learning and the joint approach, between teachers and their pupils, in making conclusions of lessons. These forms of interaction and co-planning were found to be motivating for both teachers and pupils. Teachers reported their willingness to continue this way of teaching. The following changes in teachers' practices were observed and recorded:
- Teachers started to involve pupils more in active preparation for the lessons using a partnership approach.
- Teachers organised the classroom learning environment such that pupils were very active during the entire lesson.
- Teacher made conclusions to lessons in active cooperation and involvement with their pupils.
- Teachers used ICT more actively as an integral part of the learning process.

Mentoring in Czech GIMMS

GIMMS in the Czech Republic was guided by a number of learning theories including social learning theory, role model theory, apprenticeship model and constructivist/socio-cultural theories. For example, social learning theory maintains that human beings tend to emulate the behaviour they see in others whom they respect and admire. Both social learning theory and an apprenticeship model of mentoring were identified as theories underpinning GIMMS Czech Republic. Such perspectives maintain that learning is an active construction, as well as, a process of enculturation.

The Czech GIMMS project mostly used an *apprenticeship model of mentoring* implying master teacher or teacher educator with particular skills to be passed on to the novice teacher or pupil. There are clearly other dimensions to this particular mentoring role, such as passing on an understanding of the purpose and place of a number of skills and, more controversially, assisting with the expectation that the new teacher needs to acquire particular professional values and attitudes within the workplace setting, such as the classroom or school. Both of these dimensions – skills development and values development – demand elements of 'role-modelling' in addition to apprenticeship as an important practice based element of teaching and teacher education.

The apprenticeship model of mentoring adopted had a special form. Here, the mentor was a teacher, who watched over mentees (students / pupils), through teaching a range of new topics constructively in physics, mathematics and computer science. The teacher cooperated with their students (in this case, pupils) and actively elicited their models of understanding and corrected any misconceptions. This model of mentoring was designed to enable learning to occur for the mentor and mentee through observation, socialisation and enculturation. This form of mentoring as contextualised in schools was reflective but not reform-oriented and differed, in this latter regard, from models espoused by the GIMMS team.

The mentors reported that positive outcomes from the GIMMS project for them included reflection, professional development, personal satisfaction and interpersonal skill development. Two problems reported as common to all mentors were found to be lack of designated time for this mentoring work and lack of training. It is interesting to note that the latter is consistent with requirements for a transmission model of teaching, learning and mentoring. Lack of time was the most frequently cited issue. Other problems included negative mentee attitude, pressure, and professional experience/personality mismatch.

Positive outcomes, from GIMMS Czech Republic, for the *mentee* included personal satisfaction, coaching ideas, challenging assignments, and access to 'new' and innovative curriculum resources. It appeared that co-educational groups of girls and boys were more active and achieved better results than single sex groupings. It is to this aspect of gender awareness that we now turn.

Gender awareness in the Czech GIMMS project

Resources with many practical tasks and examples from real life to positively motivate students and pupils in their study of mathematics and science were collected. It also attempted to have these resources and practices gender-inclusive. This was done through proposing ways to generate equal opportunities for girls and boys across the sciences and mathematics through using materials and approaches that would motivate both boys and girls in their study of these subjects. Therefore, gender awareness was interpreted within the study as providing equal opportunities for boys and girls. It did not take into account the role of the teacher in their classroom practices or their conceptual understandings.

Teachers in the study noted a number of gender based orientations in their observations. The observations and perceptions of teachers fell under the following three themes: willingness to explore and experiment; achievement in tests and general observation. For example, teachers observed that boys were generally more adventurous and experimental in their exploration of software and use of ICT, in physics, mathematics and computer science. While it appeared that boys in the study were more accurate in calculations in mathematics it appeared that it was girls in physics that were better prepared for testing and examination. Girls also appeared to better manage the theoretical aspects of computer science, with boys appeared more interested in real world applications. Overall in mathematics observations revealed that co-educational groups of boys and girls were more active in discussion than some single sex groupings in the study.

Dissemination of GIMMS Czech Republic

Project results were disseminated to experienced teachers, student teachers and teacher educators in a variety of ways over the lifetime of the project. CDs of the resource materials in the project were distributed to interested teachers and at a number of university seminars for teachers and student teachers that were held during the project. Findings were presented at the *European Science Education Research Association Conference,* ESERA 2009 in Istanbul (Turkey), held during August 31st – September 4th 2009.

Implications and findings from the Czech GIMMS project

In the Czech Republic generally university cooperation with primary and secondary schools is perceived as important because of better promotion of science, mathematics and computer science and for the exploration of better pedagogical innovation. For example, university lecturers often work with talented students in arranging, consulting and evaluating competitions. This cooperation is for students and pupils at secondary and primary schools not only important for better understanding of the problem, but also a motivating stimulus for further study. Organization of short training courses at university for students from secondary schools occurs often. Participants meet with experts and new knowledge developed in the field by the university is shared. Students not only acquire new knowledge, but they also increase their overview of the field of their interest and may become motivated for further study. For example, computer science courses organized by the University of Ostrava are widely popular among students in the local region. Not only students, but also their teachers are interested in these courses. Teachers at secondary schools expect from the courses they attend at the university to gain new motivation and inspiration, and to be refreshed, for their classroom teaching and lesson preparation.

Czech GIMMS was an example of one such school-university partnership approach, between four schools, eight teachers and 157 pupils to explore innovation and change. Teachers worked alongside teacher educators and researchers to assist the development of constructivist pupil-centred approaches in teaching and learning of science and mathematics. The framework for the project involved a broader conception of mentoring and inclusion than was interpreted at local level. This introduced some contextual constraints into the project.

Interpretation of innovation, mentoring and gender were indeed understood in GIMMS to be dependent on national and cultural contexts. Results from PISA 2006 had positioned the Czech Republic above average in both science and mathematics domains. The cultural context at schools showed extensive use of transmission models of teaching although the impetus for newer and more innovative approaches was being supported by the introduction of new dynamic interactive technologies. GIMMS Czech Republic challenged the cultural background and context of traditional teaching in case study schools and teacher-centred education generally.

Findings from Czech GIMMS showed that teachers did experiment with activity based learning, with different forms of technology, and with newer ways of pre-planning, co-planning and evaluating lessons. However, mentor-mentee relationships were based more on novice-expert relationships of learning than co-inquiry and reflective approaches.

The issue of time for these new approaches remained a consistent constraint throughout the project. Gender was interpreted as about the behaviour of pupils rather than in broader terms of the continuing awareness of pupils differing needs or the changes required in the practices and conceptual understandings of the teachers or teacher educators.

The cultural and pedagogical shift in emphasis from teacher-centred teaching to improving pupils' motivation and actions is needed to ensure the preparation of future teachers and the continuing education of existing teachers. This has policy implications for the continuing education of all teachers, and science and mathematics teachers in particular. Making space for practical skills and competence training in modern teaching methods and application of information and communication technologies is a necessary beginning in this direction.

Projects such as Czech GIMMS, show the types of learning outcomes in innovation and pedagogical change that can be obtained from dedicated research and development school-university partnerships of this type.

Passive transmission models of teaching clearly have no place in today's classroom. There is a need for greater emphasis on interactive learning and relationships between teacher educators, teachers and pupils. Use of activity learning and ICT-enhanced learning are known to support broader and deeper learning and thinking processes. Findings from Czech GIMMS, as one exploratory research and development project, nesting within six other European case studies has implications for the education not only for future teachers, but also for the continuing education of existing active and experienced teachers.

B) Analytical case study

Chapter 9:
Germany II: case study design and results about innovation, gender and mentoring

Manfred Lang and Gunnar Friege
IPN, Leibniz Institute for Science Education at the University of Kiel, Germany

How to realize educational innovation, change and mentoring

Schools not only in Germany but all over Europe are experiencing a rapid change of educational practice in accordance with reform requirements and societal demands. In order to guide and develop this process, modern and innovative solutions for mentoring across the full professional lifespan of teachers in teacher education need to be found and realised. Competitive mentoring models need to be studied in more detail in order to find adequate and appropriate strategies for educational innovation and change in teaching and teacher continuing education in the school setting.

In the case of Germany teacher education and mentoring in school environments are seen as a central effort to support innovative processes. This conception of mentoring and teacher continuing education is based on modern insights from research of curriculum innovation, taking into account the complexity of educational systems. One such report is an international OECD study of innovation in the field of science, mathematics and technology education (James, Eikelhof, Gaskell, Olson, Raizen & Saez, 1997). In this study the authors come to the central conclusion that a more broad-based and extended view of teacher professionalism is required:

> "Successful curriculum innovation and consolidation is dependent on a more thorough-going and comprehensive view of teacher professionalism" (James et al., 1997, p. 483).

Their view of curriculum innovation is systemic and uses the metaphor of an ecosystem, within which innovation in the school or classroom is a culture, with the teacher at the heart of this culture. They reject a common systemic view from mechanical engineering, where change in one part of the system would have to be matched by consequential changes elsewhere. As one of the authors of the OECD study Olson (2002) notes, most often this mechanical system leads to the roll-out of policy where teachers are left out of the power process neglecting feedback and insights that might come from teachers. In addition he points out, that the system of teacher professional supervision does not support the development of professional autonomy and collaboration of teachers. The function of teacher supervision is perceived as part of this reductionist mechanical system assessing success only in terms of implementing system-wide changes and outcomes. A way out of this dilemma might be to support teachers in collaborative mentoring and reflective practice for systemic reform: "These professional practices entail questions about human values, beliefs and moral considerations – questions about tradition. Teachers need to be able to bring such issues to the policy table in the context of systemic reform" (Olson, 2002).

The concept of mentoring studied in GIMMS Germany II involved the introduction of beginning teachers into school practice. There are different perspectives of mentoring represented in models such as described by Maynard & Furlong (1995). They distinguish the apprenticeship, competency and reflective model. At present the reflective model is of special interest for educational innovation and change in educational systems. A reflective mentoring model supports a partnership based collaboration of induction. Within this partnership experienced in-career teachers act as mentors and beginning teachers as mentees within a community of learners at school, exploring practical experiences in class within a flat rather than a hierarchical structure. After decades of individualism in teaching it is now realized that teachers are more effective when they can learn from and are supported by a strong community of colleagues. The literature also argues that there can be joint learning for both mentor and mentee in this type of reciprocal learning – beginning teachers can benefit from mentors just as much as mentors can benefit from dialogue and exchange with beginning teachers (Hargreaves & Fullan, 2000).

As opposed to a reflective mentoring model an apprenticeship model of training defines skills, competences and knowledge imparted to the trainee by a master teacher. It is more of a top-down approach. It assumes that all professional practice by experienced teachers is 'good' practice and worth emulating and it also retains strong divisions of power between those that 'know' and those that don't 'know'. In GIMMS Germany II these competing models of mentoring are expected to contribute differently to innovation. How do they fit to mentoring approaches practiced in different educational systems of German states? This is the question we posed in our analysis of mentoring of beginning teachers, with different case studies, in two distinct regions in Germany.

Different case studies

The starting point for the inquiry in the GIMMS Germany II case studies was with beginning teachers in the second phase of their teacher education preparation, taking place at the school site. Depending on the systemic approach taken there can be a maximum of difference in crucial aspects of teaching and teacher education with regard to autonomy and collaboration in a community of beginning teachers, experienced teachers and mentors. This is presumably the case with school systems in different federal states that take divergent systems and models of mentoring as a basis for teacher education and teaching in the preparatory service.

One system to be explored with a case study school was in Schleswig-Holstein, a federal state that emphasized the role of schools as partners in teacher preparatory service. The second system was that of Lower Saxony, as described in a number of policy documents. As opposed to Schleswig-Holstein Lower Saxony clings to a more traditional apprenticeship model of teaching, teacher education and mentoring. Both systems refer to the education, training and mentoring of beginning teachers for the upper secondary level. We will first consider the education, training and mentoring of teachers generally in Germany as background and context for this study.

Preparatory teacher education for secondary schools in Germany

In Germany the process of entering a teacher career has two phases. In the first phase the student teacher is engaged in university courses covering two subjects and subject related education together with pedagogy. The university courses are part of academic disciplinary learning. They are modular in design and open for diverse areas of vocation. Academic studies of this type lead to a master's degree or a first teacher examination performed by federal state examination agencies. The second phase is a more practically oriented preparatory service. This practically based part generally takes two years and has different emphases depending on the federal state and the type of teaching career. On the one hand it involves sitting in on lessons together with both guided and independent teaching at regular schools. In this way mentoring is partially supporting the beginning teachers' learning process. On the other hand the practical training at school is supplemented by studies in educational theory and subject-related education at teacher training institutes which are meant to reappraise and consolidate practical experiences. The preparatory service concludes with a second teacher examination. This is the prerequisite for a permanent teaching position as an employee or civil servant.

The preparatory service for beginning teachers or "Referendariat" is a second phase of teacher education after the first state examination. It generally lasts for two years and starts at the beginning of the school year or half a year later. Beginning teachers are assigned to a school where they usually remain until the end of their teacher education. This second phase of teacher education varies between the federal states. After successfully completing the second phase, of the so-called Second Examination, it is possible to change the federal state. Differences between the federal states consist, among others, of the amount of teaching a student teacher has to do independently at the school, the number of lessons to observe, the amount and kind of teaching courses to be taken and the period of time for this phase. Student teachers in this phase have different people guiding them as mentors.

Two experienced teachers, usually referred to as mentors, accompany the student teacher in each of the two selected subjects. The courses for student teachers are usually held in groups outside of the school and are given by experienced teachers who automatically assume mentoring functions. These courses are given for both the education aspect and for the individual subject specialisms.

It should be noted that the second phase of teacher education is currently in transition in many states and the findings in this study are seen as more of a snapshot in a changing time in teacher education. It is anticipated that reform measures in the two federal states, which are considered here, will continue and evolve in the coming years and has already been changed since this project concluded in 2009. One aspect under consideration is the shortening of the second phase, while strengthening the integration of classroom activities (practical aspects) in the first phase of teacher education and training. Despite considerable differences and widely complained problems, mostly concerning mentoring in the second phase of teacher education and training, this is the baseline of teacher education in the federal states all over Germany.

An attempt to discuss the different mentoring approaches of the German states and their problems in a framework is offered by the German ministry of education conference (KMK, 2009). They provide a competency approach to advisors in teacher education (KMK 2009, http://www.kmk-format.de/material/Beratung/Handreichung_201108.pdf). They call people responsible for teacher education, in the second phase, advisors as a comprehensive term used in a variety of ways in different states for mentors, tutors, supervisors, moderators or others. At the centre of this standards approach are competency profiles for advisors and mentor teachers. These competencies are arranged in a matrix with dimensions *knowledge/development* with regard to subjects, system, group, self and *action* about inquiry, application, communication and evaluation. (http://www.kmk-format.de/material/Beratung/Kompetenzmatrix_BfU.pdf)

This case study was concerned with the second phase at schools, for beginning teachers. It is an exploratory study comparing the different mentoring of beginning teachers offered in the states Schleswig-Holstein and Lower Saxony. The development of teacher education in Schleswig-Holstein is of special importance, because the preparatory service for beginning teachers was revised in 2004 applying a standards approach (Christensen, Glindemann & Riecke-Baulecke, 2008). In this part we only refer to the comparative case study of different states. Before getting into the comparative case study details we need to know some basic structures and background of the teacher education systems for beginning teachers in the two German states of our inquiry.

Beginning teachers in secondary schools of Schleswig-Holstein

Schleswig-Holstein started a comprehensive reform for the second phase of teacher education and training including mentoring of trainees, but also aspects such as the marking process. Beginning teachers are currently allocated to training schools teaching on their own for about 10 hours per week. At the schools they are assigned a mentor in each of their subjects. The mentor teacher gets a reduction of two hours of lesson requirements and is invited to participate in class regularly and support the beginning teacher, as mentee, in their preparation of curriculum, teaching, learning and assessment. In addition, there is a coordinator who specifically serves as a school mentor. This coordinator will convene regular meetings for discussion of broader themes, such as internal differentiation in teaching or issues such as discipline. Demonstration lessons at the school are a key feature of this reform. These are attended by the school principal, beginning teachers as mentees, experienced teachers as mentors of the beginning teacher and the school coordinator. After a demonstration lesson a collaborative reflection takes place, which lasts at least an hour.

The theoretical part of the preparatory service takes place in so-called modules delivered by module-suppliers. These are eight-hour modules. Beginning teachers are supposed to choose 45 pieces during the course of their education. The modules take place in different locations so that the beginning teachers must accept having to travel some distance in certain cases. Two of these modules are expected to serve for examination papers in each subject. This work is conducted in cooperation with the respective module supplier, who also corrects and marks it. A second correction is not scheduled. In the course of the examination paper the module supplier comes to the school for an hour of a corresponding lesson. As a rule, this is the only opportunity to get feedback from a module supplier about the given lesson. The examination is conducted by two module suppliers and the headmaster. The mentors, mentor teachers at the school, do not have any formal part on this marking process but may be present during the meeting and give comments at the end.

Beginning teachers in secondary schools of Lower Saxony

The preparatory service for beginning teachers in secondary schools or "Referendariat" lasts in Lower Saxony for two years or one and a half years. This may be less if sixteen weeks practical training or attendance at an academic exchange is counted for credit. The beginning teachers are assigned to one school accredited for the purpose of teacher education.

Beginning teachers are familiarized with the teaching methods by sitting in on classes during the first three months. After that they conduct autonomous lessons, in two or three subject areas, without the presence of another teacher – typically teaching for eight hours per week two subjects or twelve hours per week for three subjects. Soon after they start at the school they begin planning and delivering classroom lessons for four hours under supervision of the mentor teacher. Regularly every two to three weeks the beginning teacher invites the educational instructor, the subject specialist instructor or the education coordinator of the school to attend a lesson.

During the first year beginning teachers have to write an assignment referring to two lessons. In addition, they are expected to invite all instructors to two "special lesson visits" per subject in which the visitors critically evaluate the lesson. For this an outline with about six hours of lessons must be submitted describing and justifying the sequence of the lessons, selection of groups for learning, didactic construction and choice of methods of instruction and learning. The theoretical education about pedagogy and the subjects is carried out in seminars every week in the afternoon. The phase of autonomous teaching ends with the reports and evaluation of the instructors and the school headship. If the grade is at least "sufficient", the student teacher is then put forward for the 'examination lesson' which takes place in the last quarter of the school year. This typically takes one hour per subject. The educational instructor, the subject specialist instructor, a representative of the school headship and a schools inspector of the state education authority watch and judge the lesson. After the last lesson examination an oral examination for one hour follows. The results are summarized as marks about the training, the assignment, oral examination and the two examined 'special lesson visits'.

Elaborating research questions about innovation, gender and mentoring

The GIMMS project had as one of its focuses the differences and similarities of mentoring models including the inclusion issue of gender in the participating European countries. GIMMS Germany II used these insights to develop and pilot mentoring relationships between induction and in-career teachers in aspects of physics, chemistry, biology, and mathematics. Hence one of the key research questions to drive the project is: How can we develop better partnerships between initial, induction and in-career teachers for continuing professional development?

The work with beginning teachers, in the second phase of teacher education, was about a comparison of different mentoring models realized in different German states. One state (Lower Saxony) emphasizes the apprenticeship approach to mentoring, teaching and teacher education. The other state (Schleswig-Holstein) is moving toward a more collaborative reflective model. For comparative purposes a description about the apprenticeship approach in Lower Saxony, gleaned from policy documents analysis, was contrasted with a report and outcomes from questionnaires and interviews from beginning teachers and a mentor from one school about the more reflective approach in Schleswig-Holstein.

The idea behind the Schleswig-Holstein model was to get away from the old traditional model where only one expert teacher was a contact person for beginning teachers. The new approach in Schleswig-Holstein is to use training modules offered by the state institute for teacher education IQSH. Beginning teachers have to learn about these modules and the person responsible for the module in the IQSH is responsible for the students' learning and support during their practice placement at the school site. Parallel to this a beginning teacher can cooperate with one or two mentors at school, depending on the subjects.

The thinking and rationale behind the induction programme in Schleswig-Holstein, with modules and mentors, is for beginning teachers to become more reflective in teacher education and innovation. Questions arise, how this model in contrast to the model in Lower Saxony supports reflective mentoring, how mentors are qualified to participate in reflective practice and what kind of background beginning teachers were bringing to the school as reflective practitioners and partners in learning, innovation and inclusion.

Beginning teachers, in the second phase of education, were found to be overloaded with preparatory work and examinations. They also experienced gender differences in their social interactions as male learners and female learners confronted with mentors or teacher educators. This is the gender-awareness perspective of the beginning teachers at the training school, and is a different perspective of gender sensitivity in the school community as opposed to gender awareness in classroom teaching.

Method of tracing innovative aspects of mentoring and gender differences

Six beginning teachers and one mentor teacher in an upper secondary school in Kiel, Schleswig-Holstein, were asked to participate in the study during the time period of August 2008 and November 2009. The mentor teacher was asked in an interview about the newly introduced official teacher education concept and its adaptation at school, the main tasks of the mentor teacher and the coordinator at the school, the training and certification of mentors and consideration of gender aspects in the school. In addition the mentor teacher gathered policy documents about state requirements for the certification of teachers, the school concept about teacher education and guidelines from the coordinator.

In 2009 the second phase of the research study took place. This second phase of data collection involved questionnaires and semi structured interviews for beginning teachers. The questionnaires consisted of six parts:
1. Importance of topics about beginning teacher education in the school
2. Support of beginning teachers' work
3. Collaboration in school especially between beginning teachers and mentors
4. Autonomy and workload in daily practice
5. Experienced gender specific differences in teaching and mentoring
6. Factors which influence the success or failure of beginning teachers.

In addition, beginning teachers were interviewed about their experience with the teacher education system, and the mentoring process, including gender aspects as part of it. The results from this inquiry, about aspects of teacher education and mentoring in Schleswig-Holstein, were compared and contrasted with a critical report about similar aspects from Lower Saxony.

Results and findings

The answers of the *mentor* in the interview can be taken as a basic evaluation of the teacher education system for beginning teachers in Schleswig-Holstein. They offer a general perspective of the newly introduced dual system of theoretical education and training with standards based modules offered by the state institute IQSH and practical training with mentoring support in selected schools: Beginning teachers can choose modules in different subjects and educational topics, produce a portfolio and write an assignment about it. Different personnel from the IQSH present these modules in one-day courses in a variety of different geographical locations. Mentors are not always acquainted with the new modules in advance, but they are cooperating with beginning teachers in the co-planning and preparation of lessons, more often on a number of different levels such as offering practical advice and teaching outlines or discussing typical beginning teachers' mistakes.

One problem identified with this system of modules is seen in the continuing change of presenters and examiners. Beginning teachers, in the newer system, do not have only one person for reference with publicly accountable expectations for their trial lessons and final examinations. They receive feedback instead from a group. In this regard the old system, with study guides as mentors in fixed groupings, was perceived by some student teachers as being more favorable. In addition presenters of modules at the education institution and mentor teachers at the school do not have the same background and competencies. Mentors only receive a general preparatory introduction into their tasks. These tasks are not related to specific goals for preparation of beginning teachers. A more specific and targeted education for mentors is judged to be more favourable.

Beginning teachers answers in questionnaires and interviews partially reflected these critical points of missing coherence between theory in modules and school practice and contradictory demands for examinations. In the questionnaire they agreed that the modules for their subjects were helpful and that they got sufficient support and time for lesson preparation from the school. They did not find help in the modules about pedagogy and did not have enough time to cooperate sufficiently with the mentor teacher at the school or to discuss their work or experiences sufficiently with other beginning teachers. Content of subjects and its theory in modules seem to be separated from school practice. It clearly needs to be elaborated in congruence with mentor teachers. This point is explicitly made in one interview [4]:

> "This is absolutely separate. There is something demanded from us by the IQSH [State Institute for Quality Insurance in Schleswig-Holstein] and what we are discussing there with other beginning teachers. And then there is the school, what happens there. And this is completely separated, which is a problem. On the one side we have to do what the module presenter is demanding from us, starting from educational standards, competencies and goals and on the other side in the school they say – pfff – curriculum, educational standards. This is mostly of no interest; they do not know what a curriculum is".

Beginning teachers marked in the questionnaire that they missed to a high extent that the module presenters and others as examiners explained the criteria for the examination and that they did not get sufficient help for the examination paper. Concerning the demands from module presenters, mentors and the headmaster they feel that they have had to serve different masters with many opposing requirements.

The standards based modules are usually not well known by mentor teachers or principals at the school. This became a problem of agreement between this group of people responsible for the preparation and certification of examinations. In addition different modules may be offered by a variety of presenters having different ideas and ideologies about teaching and education principles. These education and teaching principles might then be favoured differently by mentors as reported [4]:

> "The module presenters are not in agreement with their requirements at the end, although there are standards for that and information for preparation defining the criteria for correcting our exam home work … . One time it was required that I show my professional development as a teacher, what I have learnt, how did I manage the lesson and how I profited from that personally. In the other homework I had to show how I supported the pupils, how did I use different criteria for the development of pupils? These are completely different tasks … . This is the point with the mentor that we shall do what we are required but he is not interested in educational standards or methods, there is only interest in subject matter, subject matter, subject matter".

A possible solution to bring greater levels of coherence to these divergent views might be to have a better reflective collaboration between the various actors as suggested in the interview [2]:

> "I have experience about better exchange, defining the other mentor type. I always found out that mentors liked to get material we brought from the modules … . And I believe that it is meaningful that mentors and beginning teachers go together to module meetings. This I have done once with a Math mentor. This is something different, if you meet and can talk together about it".

Collaboration, reflection and discourse, in the larger more extended professional network, between the module presenter as teacher educator, beginning teacher and mentor teacher gives a better chance of specific learning for all while linking theory of modules with the practical context of the classroom and school. This need for specific learning was already mentioned in the mentor interview as a problem of preparatory courses for mentors. In interview [6] a beginning teacher is confirming this complaint about a lack of specific preparation of mentor teachers:

> "In general mentors are not well trained. This is obvious … we as beginning teachers experience, that we are trained very well for the module and then we give our knowledge to the mentors … They don't know much and in our meetings for consultation the mentors are those, who can say the least about the given lesson. There should be done more so that trainers become trainers, which are now simple teachers who guide us – nothing more".

To a certain extent a culture of cooperation and autonomy seems to be fostered through partnership based mentoring. Especially the beginning teachers from Schleswig-Holstein report a relatively high degree of autonomy and satisfaction about cooperation with colleagues in general. But the opportunities for beginning teachers to discuss their work with mentor teachers were estimated to be low. Nonetheless it was lower-order cooperation on the practical and technical aspects of professional work that seemed to dominate mentoring processes under different mentoring systems, the traditional mentoring model and the new mentoring model in the state of Schleswig-Holstein and between the states of our study.

From GIMMS Germany II research findings, beginning teachers in Schleswig-Holstein seem to experience a high degree of support in their school. Nonetheless there is very little cooperation between initial teachers, beginning teachers and experienced teachers, and teacher educators and module providers other than with their specifically nominated mentor teacher. The partnership model of mentoring favoured in Schleswig-Holstein does not seem to influence or even foster a culture of cooperation beyond the beginning teacher – mentor teacher relationship. The beginning teachers report very little collaboration within the full community of teachers at the school. There were few reported examples of higher order professional cooperation. Discussions on teaching experiences and lesson plans were scarce. Cooperative lesson planning rarely ever happens.

Beginning teachers personally experienced gender differences during their social interactions with the range of actors in their teacher education. GIMMS Germany II asked a key question in this regard: How are male and female beginning teachers treated differently? The beginning teaches, especially the female beginning teachers in our study, experienced a difference. They were complaining that their gender made a difference in discussions as pointed out in interview [3]:

> "I experience a higher richness, when men are sitting with us in module sessions. This brings more objectivity in or another point of view, which is not too female".

Female beginning teachers said that they were treated differently in some ways. When there were only female beginning teachers together, discussions were completely different in the sense that there was no rigorous and a less aggressive demand. A female beginning teacher talked about a "good girl model". By this she meant that mentors didn't challenge the female beginning teachers, confirming that they are right without going deeper into the discussion and without requiring depth of analysis and reflection. They did not like this because they wanted to learn more. If a male beginning student was present it was assumed that they would participate in a more aggressive discussion. This is described in interview [6]:

> "There was an attempt at first to start a feminine gear of fibre softener and when the women did not accept, it got really hard, very personal, very violating" ... The kind of consultation after trial lessons changed. It became extreme critical and personal, when we were together ... Men were treated harder than women. Women got an extra charm credit".

Reflection is needed to move to some new places and new perspectives. This is a central requirement for changing to the 'new' of innovation in teaching and teacher education and mentoring. Challenge is required and this was missing in the case of the "good girls' model". Beginning teachers did not give time to planning gender-aware materials for their classrooms. They had gender specific problems themselves and the pressure of finishing their examinations. But there was some consciousness with beginning teachers that they themselves are partially contributing to gender differences. In the interview [6] it was said, that

> "I believe that it [the gender difference] has much to do with the teacher herself. Some of them can better handle girls, others better boys and this happens and then it makes no difference what you are teaching. ... If I present creative tasks, girls better attend to it, but it does not mean that they can do it better than boys. If the boys give attention to it they can do it as good as girls".

From this perspective it is being claimed that the teaching materials themselves do not necessarily produce the gender differences, but the teachers can stimulate the attention of the male pupils or female pupils given the materials. Gender sensitive materials might be linked by purpose to specific topics as in the other German case study about *Biology in Context* in Chapter 5, where the teacher educator and the mentor teachers guided the attention to ethical discussions for both, male and female pupils.

Conclusions and policy implications

All in all the preparatory phase in teacher education is highly demanding for beginning teachers independent of the approach, either traditional or innovative as in the two federal states in our study. In both cases beginning teachers experienced a high workload which was challenging with respect to time management, coping strategies for frustration and endurance. The challenges were not in the demands to learn teaching for beginning teachers but to manage the different problems of the diverse approaches and divergences of demands.

The traditional approach of study seminars in Lower Saxony, and as well the same approach used in Schleswig-Holstein in the past, is at first glance an apprenticeship model with the advantage of a steady person of reference, the study seminar guide, and a coherent group. This steadiness is an advantage favoured by beginning teachers for preparation of their examinations. However, this system has the problem that a negative choice of a study seminar guide cannot easily be corrected and that the amount of contacts, and therefore perspectives on teaching and teacher education, become limited.

On the other hand the newly introduced dual based system of induction for beginning teachers, school based mentoring and module oriented second phase of teacher education, offers many contacts and challenges. It has the built in crossing of boundaries with the risk of boundaries between and within subsystems such as school, state institutes for teacher education and university. However as currently operationalized the many partners in this model, did not offer challenge as an educational innovation and change mode of reflective teacher education, teaching and mentoring. As identified in our inquiry, beginning teachers have to serve several masters with different demands until their final examination: different module presenters, mentors for different subjects, a principal, a coordinator, all not necessarily well interlinked. This might be due to the new approach with modules. It is not only a structural problem. There is a general problem of educational innovation and change within modules. It shows in the difficulty in interlinking theory and practice. Standards based modules are introduced in school without a school based systemic change toward standards in education. Beginning teachers, mentors and other teachers in school are not familiar with these standards or do not understand the difference between traditional education and standards education. Experienced teachers cling to their subjects and reduce teaching to transmission of subject matter knowledge without standards in mind.

A special problem of boundaries was found with gender differences in meetings with beginning teachers. Female beginning teachers felt treated differently in discussion than their male colleagues. The boundaries changed as soon as male beginning teachers took part in the discussion. These problems with different boundaries could be treated in a more negotiated and aware approach through a collaborative meeting within an agreed framework, of beginning teacher, mentors and module presenters as suggested in one interview. A framework for boundary crossing in activity theory might be helpful (Lang, 2009, 2012; Engeström, 2000). To cross boundaries of different communities in the school system collaborative settings are needed. They can take different appearances such as a round table, an open forum for face-to-face interaction, a *Learning Studio*, a shared virtual space for *Computer Supported Collaborative Learning* or a *Curriculum Workshop* (Lang, Couso, Elster, Klinger, Mooney Simmie, G. & Szybek, 2007). In these settings different demands can be identified, standards or modules and lesson preparations adjusted and the roles of mentors, principals, coordinators and module presenters integrated. Mentors would, in this approach, learn specific educative competencies in collaboration such as reflective school practice which is different from the current general and technical requirements for their certification.

Boundary crossing in a collaborative setting is a broad systemic approach for educational innovation and change with many actors. In this approach teachers at schools are at the centre of the setting. It is not a reductionist mechanical systemic approach as outlined in the introduction that leads to the roll-out of policy of one state, in a top-down reform approach, where teachers are left out of the power process. But limits of autonomy have to be tested.

The state governments in Schleswig-Holstein and Lower Saxony decided to give schools more freedom and self-responsibility to develop their educational practice. As a result schools in these states developed different school-based mentoring systems with different opportunities and challenges. They promoted innovation to a different degree with teacher professional supervision as mentoring that supports professional autonomy and critical feedback of teachers to a certain extent. This professional autonomy depends in Lower Saxony mainly on the mentor-mentee working alone. In Schleswig-Holstein it depends on the systemic interaction between people from the State Institute IQSH, mentors, principals, coordinators and teachers. Modules for teacher education from the IQSH were designed as a support for beginning teachers to reach educational standards and not as a political instrument for implementation. This meant that reflection was promoted within limits defined by the standards and related competencies giving educational innovation and change a framework.

The fact that beginning teachers were usually not familiar with the standards, and experienced teachers in this GIMMS Germany II study did not care about them means that it will be difficult for school based innovative decisions to be realised and successful in public education going forward. Discussing educational innovation and change in schools beyond this frame of standards will be a stony pathway, because there will be significant resistance from state governments toward localized or school-based collaborative decision-making powers or toward investment in border-crossing networks for sustainable change.

GIMMS Germany II presents evidence for the need for an extended deliberative discourse for beginning teachers that has agreements, negotiations, opportunities and challenges for all teachers, across their professional lifespan, in collaboration with teacher educators and policymakers for the improvement of pupil learning, educational innovation and change.

Part III: Cross-Case Analysis, Findings and Implications

> Teaching is increasingly seen as a professional activity requiring a careful analysis of each situation, choice of objectives, development and monitoring of suitable learning opportunities, evaluation of their impact on students' achievement, responsiveness to students' learning needs and a personal or collective reflection on the whole process. As professionals, teachers are expected to act as researchers and problem-solvers, reflecting on their own practice and assuming greater responsibility for their own professional development. (OECD Teachers Matter 2005, p. 99)

PISA findings and the findings from OECD *Teachers Matter* 2005 showed that educational innovation and change were central concerns in both teaching and teacher education. However these international findings were not sufficiently deep to get a real world understanding of the contextual and cultural layers that oftentimes constrain as well as support educational innovation and change in the school setting.

GIMMS as a research and development project was established within an agreed theoretical framework and was guided by a number of evolving democratic principles. It took the cultural and contextual landscape of teachers and teacher educators, and others into account in seven different case studies across six European countries.

The cross-case analysis from each of the national case study reports that follows explores the comparative qualitative data generated by the national coordinators in seeking a deeper understanding of educational innovation and change at the level of the school and the classroom within a changing Europe.

Findings from this analysis show that responsibility for educational innovation and change, and the public space necessary for this discourse, cannot be left as the sole responsibility of the school community or for teacher professional networks. A deliberative discourse, within an agreed democratic framework, and including a number of actors including teachers, at all stages of their professional lifespan, teacher educators and others, such policymakers, is required for teachers and teacher educators to collaboratively and reflectively develop educational innovation and change.

This border-crossing deliberative discourse when conducted within a specially designated incubation space has the capacity to challenge traditional schooling and teacher education. It does this through offering the appropriate level of support and pressure to teachers while guiding innovation through access to the latest research and subject knowledge findings.

The educational innovation and change process as exemplified by the GIMMS studies, in the real world setting of the school, and beyond the school is shown to be a complex process that requires a range of systemic supports and structures. It offers a changed and new role for teachers, teacher educators and others, including policy-makers.

Chapter 10:
Educational innovation and change as deliberative discourse across borders of educational systems in Europe: a cross-national analysis of GIMMS findings

Geraldine Mooney Simmie and Manfred Lang
University of Limerick, Ireland and IPN Kiel, Germany

This chapter considers the background and context for the study and the evolving theoretical framework that regarded educational innovation and change as pedagogical and political text (Aronowitz & Giroux, 1991; Pinar, Reynolds, Slattery, & Taubman, 2008; McLaren, 1991a and Mooney Simmie & Lang (in preparation)). Mentoring relationships between teachers and others within the extended discourse of the project were set within a platform of democratic values and principles (Mooney Simmie & Moles, 2011). Using this framework the cross-national qualitative research findings in the form of transcripts from interviews and team meetings, questionnaires and reflective journal entries are then analysed. Findings indicate that for educational innovation and change to be realised a number of sustainable 'structures', 'supports' 'thinking time' and 'public spaces' that cross borders need to be provided. In the case of GIMMS these border crossings included national borders as well as institutional borders (Hansen, Gräber, & Lang, 2012). In this way teachers, across their professional lifespan, were continually engaged in a deliberative discourse with teacher educators and other stakeholders, such as policymakers.

Context for the study

Educational innovation and change in teaching and teacher education does not readily happen in stable school cultures with their traditionally inherited and robust capacity to resist change. GIMMS approached educational innovation and change in the school setting through the establishment of a number of different settings and groupings, 'public spaces', perceived as *incubation spaces* that crossed borders within and between schools and other institutions, including higher education institutions. The deliberative discourse within these spaces facilitated the generation of creative and critical thinking and generated a *creative dissonance* between the variety of actors over the lifetime of the project, 2006–2009. In this way capacity was built for supporting and pressuring teachers, across their professional lifespan, and assisted their development as collaborative innovators and agents of change in their classrooms and schools.

The seven GIMMS case studies, in six European countries, had a number of fiscal supports, through higher education institutes and European funding, over the three years of the project, 2006–2009. These supports included moral, material and structural supports for the setting up of regular 'public spaces' for face-to-face meetings and virtual meetings – teacher educator meetings, teacher-teacher educator meetings and teacher meetings. There were some additional policymakers-teacher educator-teacher meetings in some case studies. These meetings resulted in the development of jointly agreed strategies and instruments, for example, reflective journals and resulted in the development of a variety of pedagogical resources. GIMMS retained at all times the complexity involved in interweaving a tapestry of diverse threads – innovation, mentoring and gender awareness – that were cross-linked with threads of commonality and diversity within a variety of contexts and across six national borders across Europe.

OECD findings as a backdrop to the study

GIMMS 2006–2009 started and finished against the backdrop of two international reports into teaching and teacher education: the OECD *Programme for International Student Achievement* (PISA) 2003 to 2009 for reading, science and mathematics literacy among 15-year olds (OECD, 2004; Prenzel, Artelt, Baumert, Blum, Hammann, Klieme, & Pekrun, 2007; Klieme, Artelt, Hartig, Jude, Köller, Prenzel, Schneider & Stanat, 2010; OECD, 2010). The other report was the OECD *Teachers Matter* 2005 into recruiting, attracting and retaining effective teachers across twenty five countries. These reports raised questions about student achievement and teacher education and qualification in different European educational systems and challenged existing national policies in each of the GIMMS countries.

PISA tested 15-year olds in reading, scientific and mathematical literacy every three years. The focus was on real-world applications rather than on what could be described as 'school knowledge'. Results for mathematical literacy in PISA 2003 to PISA 2009, across the six GIMMS countries, are shown in Table 1. An average cumulative mean score for each country is presented. The positioning of countries in a ranking list with mean scores changed over time. In 2003 the Czech Republic and Denmark were above average, Austria, Ireland and Germany were on average scores and Spain was below average. Gender differences in mathematical scores, with high score point advantages for males, were found in most countries in 2003, 2006 and 2009 (OECD, 2004; Klieme et al., 2010 pp. 166-167). Over the lifetime of the GIMMS project, and in the first decade of this century, the Czech Republic, Denmark, Austria and Ireland steadily lost points in PISA.

Table 1: Mathematical literacy results for GIMMS countries in PISA 2003 to 2009.*

GIMMS country	Mean Scores		
	2003	2006	2009
Czech Republic	516	510	493
Denmark	514	513	503
Austria	506	505	496
Germany	508	504	513
Ireland	508	501	487
Spain	485	480	483

*The mean score for PISA 2003 to 2009 was close to 496 score points.

Results for scientific literacy in PISA 2003 to 2009, for each of the six GIMMS countries, are shown in Tables 2. While mean score points are a useful benchmark they conceal distributions within each country and within regions. Denmark and Spain are performing 'below average' according to the PISA benchmark. Ireland, Germany, the Czech Republic and Austria are above or on 'average'.

Trends in each country over the years indicate changes possibly due to policy changes in the educational systems. In the case of mathematical literacy GIMMS countries Czech Republic, Denmark, Ireland and Austria were losing scores, while Germany was gaining. In the Czech Republic scores declined from above-average levels in 2003 to around the OECD average in 2009. In Ireland, performance declined from around the OECD average to below average by 2009. For scientific literacy Czech Republic scores dropped from above the OECD average in 2006 to around the average. During this timeframe Germany and Denmark were gaining scores. The OECD (2010) interprets that PISA results, over a full decade, show whether school systems were becoming successful in helping students understand life and living in the real world context of a changing society.

Table 2: Scientific literacy score results for GIMMS countries in PISA 2003 to 2009.*

	Mean PISA score results		
GIMMS country	2003	2006	2009
Czech Republic	523	513	500
Ireland	505	508	508
Germany	502	516	520
Austria	491	511	494
Spain	487	488	488
Denmark	475	496	499

* The mean score for PISA 2003 to 2009 was close to 496 score points.

Concerning gender results on average in OECD countries, boys outperform girls in mathematics by 12 score points. It is important to note that in four out of the six best-performing countries, there is little or no gender difference in mathematics performance. The OECD (2010) views this as a signal to policy makers that skills in mathematics are not related to gender and that more can be done to raise girls' level of performance in mathematics. On average in OECD countries, boys and girls perform about the same in science. Especially in countries with the strongest performance in science, boys and girls generally do equally well, the same finding as in mathematics. This suggests that science is a domain where policies that focus on gender equality appear to have succeeded the most.

The structure, duration and location of teacher education programmes varied considerably across Europe. *OECD Teachers Matter* 2005 identified many similarities and differences between the six GIMMS countries with regard to initial teacher education, induction and continuing professional development. For example, initial teacher education varied from 3 years to 7 years. Germany stood out as having the longest lower threshold in this regard, 5.5 years. Countries offered either concurrent or consecutive teacher education degrees or a combination of both types. Concurrent programmes offered subject matter studies alongside pedagogy and professional studies. Consecutive programmes involved a period of post-degree study of pedagogy and professional studies. Both types of programmes were offered in Austria, the Czech Republic and Ireland. Denmark offered the concurrent programme. Germany and Spain offered the consecutive model with requirements for post-degree examination for teacher employment (Table 3).

Table 3: Pre-service lower secondary teacher education requirements, 2001.*

GIMMS (country)	Duration (years)	Consecutive /Concurrent Programmes	post-degree examination (teacher employment)
Austria	3–5.5	both types	No
Czech republic	5–7	both types	No
Denmark	4	concurrent only	No
Germany	5.5–6.5	consecutive only	Yes
Ireland	4	both types	No
Spain	4–6	consecutive only	Yes

* OECD Teachers Matter, 2005, Table 4.1, p. 106.

By the early years of this century there was increasing recognition that a separate period of induction was required for beginning teachers. This varied between three months in the case of Denmark and up to three years in the case of Germany.

All countries offered professional development opportunities for experienced teachers. These were offered as menus of courses, of varying duration, and in a variety of topics. They were generally fragmented, increasingly deregulated and open to increasing levels of competition among providers. In some instances continuing professional development was tied directly to continuing tenure and licensing of teachers. For example, in Austria teachers were required to complete a minimum of 15 hours of professional development in a school year. Standard setting for teachers required a broader consultative discourse:

> "Broad consultation with, and involvement by, the teaching profession and teacher educators in developing statements of student learning objectives and consequent teacher profiles are vital in ensuring successful implementation. Standard-setting must be seen as an iterative, and not top-down, process if it is to usefully inform the development of teacher policy". (OECD Teachers Matter 2005, p. 113)

There were multiple challenges involved, for policymakers and school administrators, to refresh teachers across their professional lifespan and interconnect them in ways that might assist them fulfil their newly extended professional roles:

> "The stages of initial teacher education, induction and professional development need to be much better interconnected to create a lifelong learning framework for teachers. Initial teacher education must not only provide sound basic training in subject-matter knowledge, pedagogy related to subjects, and general pedagogical knowledge: it also needs to develop the skills for reflective practice and research on the job". (OECD *Teachers Matter*, 2005, p. 95)

Improvement efforts in teacher professionalism initiated by PISA through programmes such as SINUS in Germany (BLK, 1997) or IMST in Austria (Krainer, Jungwirth, Kühnelt, Rauch, & Stern, 2002) were similar to approaches to mentoring in GIMMS. However while they all sought to improve science education through teacher education in collaborative school settings GIMMS went further and introduced the concept of a much broader and extended network of support for teachers, both within the school and well beyond the classroom and school.

GIMMS results were interpreted against the background of a general trend initiated by PISA. PISA results of student competencies were not interpreted as direct effects of political interventions but were related to cultural and social factors. Gender as one of these factors studied in PISA was of special importance in GIMMS. In PISA the average mathematical competency of male students was significantly higher than that of female students. This did not change during the lifetime of GIMMS 2006–2009. For example, in Ireland and the Czech Republic differences are smaller than in Germany, Denmark, Spain or Austria.

For scientific literacy the picture is different. Girls in the Czech Republic and Ireland outperform boys while in Germany, Denmark, Spain and Austria boys outperform girls (Klieme et al., 2010, pp. 166, 188). For preparation for life it is important to note there were marked differences in male and female student expectations of having a career in science (OECD, 2009a). Why is it important to look at gender differences? An important argument is that these differences point to areas where student backgrounds and characteristics significantly affect student performance and their future development. In PISA some of these differences were highlighted but without any consequences for interventions. In GIMMS these differences became a starting point for educational innovation and change.

An evolving theoretical framework

The cultural approach of using a deliberating discourse is a political approach using basic democratic values and protocols. This cultural approach for generating a deliberative discourse on curriculum and pedagogy is concerned with power of arguments and communicative competency (Habermas, 1984). It is concerned with establishing an ideal setting for voice that is not distorted by structural, power or physical forces. Any constraint identified, such as an aspect of the hidden curriculum, may itself become a topic for this discourse. It becomes a question of equality of respect and participation.

GIMMS interpreted educational innovation and change as a pedagogical and political text that brought teachers into public spaces to communicate and justify their practices and to re-conceptualise ways of thinking about and practicing teaching given the latest research findings (Pinar et al., 2008, Chapter 5). Aronowitz and Giroux (1991) emphasise the importance of crossing borders in diverse cultures as a process of generating educational innovation and change. In this way 'political' becomes viewed as social transformation and teachers engaged with educational change are considered public "transformative intellectuals" (Giroux & McLaren, 1986). Teachers acting in this way have the possibility to relate education directly to different contexts. They are being presented with a multiplicity of perspectives and have increased autonomy. Teaching as professional agency requires several types of supports from external agents also willing to immerse themselves in a similar empowering and challenging process. The creative and critical thinking and the new consciousness generated in such public *incubation spaces* goes beyond the borders of the school and the higher education institution.

GIMMS argued that educational innovation and change requires a designated public space for building the capacity for teachers to engage in a stance of questioning, exploring, experimenting associated with a border-crossing deliberative discourse. In this way teachers work alongside different actors, such as teacher educators and other teachers at different stages in their professional lifespan – student teachers, beginning teachers and accomplished colleagues. The word "political" in this sense is different from the usual meaning of the word, which more often refers to representative democracy. This needs to be emphasised in order to prevent misunderstanding. Our usual understanding of political activity involves the central role of elected politicians from different political parties. A direct *bottom-up* policy coming from engaged citizens is an exception in the classical meaning of a political system. But this latter type of *bottom-up* engagement comes close to our understanding of basic democracy in a cultural field of political discourse.

Different cultures in the border crossing process can be defined as educational communities such as physics or chemistry teachers in school departments, scientists in a chemistry laboratory or chemistry teachers in an educational research cluster. These educational communities have common rules of cooperation and values and display differences from other communities through working styles, traditions, language or habits. As a consequence the work of border crossing between different communities is difficult (Aikenhead, 1996). For example, concepts in the sciences have a socially negotiated shared meaning within a group such as a science class or a peer group. Students on the other hand live in a life-world of youth subcultures, which are different from science cultures. From this perspective, learning of science has the added difficulty of being a cultural acquisition. Aikenhead (1996) argues that student subcultures of 'potential scientists' can more efficiently cross these borders because their worlds of family and friends are more congruent with the worlds of school and science. In a similar way educational innovation and change becomes a cultural acquisition and equally involves border crossing between different communities. GIMMS posits that educational innovation and change become possible if different communities are willing to engage with shared meaning-making and exchange at their borders. It is at the edge between borders, in the space where creative dissonance resides between different communicates, that educational innovation and change are made possible (Hansen et al., 2012).

In GIMMS each participating country developed a setting, in which stakeholders from different communities developed a shared meaning about innovation, mentoring and gender in teaching and teacher education. In most cases these stakeholders came from an education department of a higher education institute, teacher education institute or university, and different science and mathematics departments of schools. The values underpinning this discourse, with associated protocols, were agreed in advance. This precaution was necessary as political, administrative stakeholders or expert scholars may otherwise dominate a deliberative setting. This needs to be acknowledged and curtailed. If changes are intended then administrative stakeholders and policymakers need to partake in a discourse to transmit decisions to the political system that can support, stall or prevent changes. Political stakeholders may be interested in a justified discourse with various stakeholders in order to have a broader basis of supportive arguments and understandings in the complex world of educational planning.

Selection of deliberative stakeholders

What criteria might be critical for a selection of deliberative stakeholders? A major point appears to be autonomy. They need to be independent from other authorities or financial constraints. For the development of educational innovation and change these chosen deliberative stakeholders cannot be neutral, they need to identify and declare their ideological position. They need to be engaged with the school and supportive to schools, teachers and students. Combinations of teacher educators and teachers, across their professional lifespan, displayed this autonomy in GIMMS when working together to progress educational innovation and change through reflective, democratic and border crossing mentoring relationships of learning and co-learning.

Ethical considerations inside a deliberative discourse

Finally, ethical questions needed to be considered for the conduct of a deliberative discourse that crosses borders. Such ethical questions include equality or flat structures and rules of communicative competency – for example, equal rights to talk and neglect of status. In GIMMS this ethical stance was underpinned by egalitarian theory (Lynch, 1999). Ethical principles guide decisions: how to offer and justify subject knowledge; to structure lessons; set expectations; and to answer questions. They are reflected in the way curriculum developers talk to others about colleagues or their school, respect traditions and enforce policies or handle school norms.

Considering the case studies, using this political lens of deliberating discourse for innovation, focuses our view on the specific borders in a fragmented school system, diverse voices that open these borders for shared meaning making, collaborative settings with different stakeholders, a diversity of mentoring relationships, deliberation of hidden constraints and democratic attempts for equality of respect in an open 'public space':

> "(…) in which various discourses and practices intersect and diverge, reflecting and creating a political location … political struggle is discursive; it involves destabilising patterns of thought which cannot finally, be separated dualistically from physical behaviour of actions." (Pinar et al., 2008, chapter 5, p. 309)

Research methodology

GIMMS, as a research and development initiative, was largely an exploratory qualitative analysis. The analysis involved two phases. In the first phase case studies, set within an agreed theoretical framework, illuminated the innovation interventions within each country. The process and product aspects of these were continually reported through minutes of team meetings and through the final reports found in Part II. The critical writings of the team were an additional data source in this phase of the study. In the second phase qualitative data was obtained across national borders through questionnaires and interviews by the authors of this chapter and validated by the team. The study included a triangulation of multiple cross-national perspectives and data sources. Data sources used in the final analysis included: questionnaires, reflective statements, interviews and transcripts from minutes of team meetings. Analysis involved a constant comparison thematic approach using the lens of a deliberative discourse, to open borders to a 'public space', for shared meaning making about educational innovation and change in teaching and teacher education.

Findings from GIMMS

GIMMS coordinators, as teacher educators, argued for a conceptual framework that gave 'voice' to teachers, across their professional lifespan. The GIMMS teacher educators were willing to 'problematise' their role in this complex process:

> "We, as teacher educators, are challenging our own thinking in this project and supporting each other to develop newer and more innovative ways to assist teachers and (pupils) become more imaginative ... we are willing to dialogue, not in a linear way, but embracing all the chaos of young people's growth and development, we are embracing the socio-cultural perspective ... the teacher of mathematics and science in the twenty first century needs to have the ability to create classroom dynamics that develop a relationship of learning with young people and the meta-cognitive skills required for ongoing self-evaluation". (GIMMS Team Meeting 04/10/2007, p. 16)

The project involved eliciting learning with teachers as public educators and professionals and facilitating them, in turn, in developing pupil-centered teaching approaches. The team perceived the change task as one of challenging traditional schooling and teaching approaches, through consideration of recent research findings. However the team resisted offering an alternative ideology and remained responsive to socio-cultural context:

> "Constructs like innovation and mentoring have their own national and local flavour and differ from country to country. The research needs to capture this difference. The need for science and mathematics education to be both con-

text-led, (values-led), and content-led is a great challenge for the project, a balance needs to be maintained." (GIMMS Team Meeting 15/10/2008, p. 34)

Educational innovation and change was clearly not about reproducing existing practices and relations in the wider educational landscape. It was much more about developing productive mentoring relations and practices that challenged tacit assumptions and routines. Innovation, taking this systemic view, is complex and requires reflective practice 'to make tacit knowledge of routines and gender gaps public for discourse' (Coordinator X Reflection 26/04/2008). In a dyad between mentor and mentee, often in a school setting, there may be no explicit impulse for teachers to go beyond existing classroom practices and routines. An extended community, that crosses boundaries, has far greater potential to move practice beyond existing models. The joining of teachers with teacher educators, in the case of GIMMS, brought the research findings and the lens of the literature to the initiative:

> "Innovation is complex and needs dynamic views of school-based teacher education, not only the single isolated school. The teacher is viewed within a network of other teachers, students, school administrators and links outside school (parents, community, extra mural learning, policy, higher education). In this systemic view changes are possible through collaboration and reflective practice in these networks and so overcome outdated traditions (hierarchical teaching styles, constraints and gender gaps)". (Coordinator X Reflection 26/04/2008)

Deliberative discourse about innovation in the GIMMS case studies

The seven case studies – Austria, Czech Republic, Denmark, Germany I und II, Ireland and Spain – were interpreted in the final analysis as political struggles in a deliberative discourse for shared meaning making of diverse voices. These struggles supported, in a variety of ways, teachers' efforts at communicating, justifying and re-conceptualising their pedagogical and professional practices.

In GIMMS *Ireland* teachers agreed to work on innovative approaches to teaching and learning science and mathematics through collaboration in a variety of 'public spaces'. This was new for schools and teachers. The teachers engaged, reluctantly at first, in reflective writing and later involved their pupils in a similar process. The OECD *Teaching and Learning International Study* (TALIS) 2009 study showed that teachers in Ireland were less familiar with higher-order cooperation that involved discussion and debate on matters of curriculum and pedagogy. GIMMS *Ireland* teachers became involved in multiple layers of collaboration, in their school setting, with teachers from other schools and with teacher educators. They discussed their progress with national policymakers (Mooney Simmie & Power, 2012). This discourse was held at a time when Ireland was preparing official guidelines and policy for teaching as a continuum, requiring teachers to become inquiry-oriented, reflective and collaborative, and a pupil-centered lower secondary education system with a different type of assessment system (Teaching Council, 2011; NCCA 2011).

GIMMS *Ireland* resulted in profound changes in collaborative and reflective practices for teachers, and their pupils, in four different school settings. It resulted in the production of a number of innovative teaching resources. While teachers engaged in higher-order collaboration the resources produced were mostly designed to assist pupils with lower order questioning, and mirrored the state examination in most cases (Mooney Simmie, 2007a). GIMMS Ireland was less successful in generating a productive discourse on gender. It was difficult to engage teachers in a discourse about gender as they mostly felt that gender was not an issue in their science teaching.

In GIMMS *Czech Republic* four case study schools focused on developing constructivist approaches to teaching physics, mathematics and computer science. This was unusual in a tradition where schools are tightly controlled, with detailed lesson plans and all topics and activities documented, on a daily basis, in a class book for public accountability to inspectors. In GIMMS *Czech Republic* teachers supported newer ways of teaching through creating 'space' for pupils to explore new topics together. Pupils reported that they enjoyed lessons using information technologies and less traditional teaching. Pupils were given some freedom to discuss their ideas and to work in groups. Innovation and change was interpreted as a move from transmission to social constructivist models of instruction. This coincided with a national curriculum reform. The project succeeded in overcoming strict boundaries between classroom and official requests. However it did not initiate a deliberating discourse about more open teaching methods in a forum with political relevance. The Czech coordinator (EV) recognized that

> "Pupils enjoy this subject and the working atmosphere was very cooperative, it was the most interesting aspect of this project".

Support for change from mentor teachers did not exist and is not usual in the educational system. Voices in a public discourse for change in teacher education were limited to experienced teachers and the coordinator. Supervisors that were controlling activities at the level of the classroom were missing in this deliberative setting. The case study generated different types of classroom organisation for teaching and learning and a wide range of resources were made available through web-pages and CDs. Mentoring followed an apprenticeship model with strict novice-expert divisions. Similar to Ireland gender awareness was narrowly interpreted with poor teacher responsiveness.

In *Spain* the level of student achievements in PISA were expected and did not initiate a radical shift in science education. PISA had positioned Spain in the 'below average' category in both science and mathematics. Usually lesson planning is limited to a formal sequencing of content without special attention to pedagogy as noted by the Spanish coordinator: "In school they discuss very little about pedagogy, they discuss about content, what are we doing now or later, what is the best book." Mostly teachers are qualified as subject matter knowledge specialists, trained in a novice-expert system as student teachers and beginning teachers, and do not generally get involved in continuing professional development. Historically teachers have enjoyed high status in secondary education as knowledgeable people. The system has many written examinations. Teachers have a high level of freedom, with low levels of public accountability at the school, and appear to engage mostly in lower order cooperation, sharing administrative matters in preference to collaborative reflection:

> "In the school they discuss very little about pedagogy they discuss about content, what are we doing now or later, what is the best book we are going to use, most of the discussion is about formal things, like, when we start, when we finish, if we do a visit outside or things like that, it is not real cooperation for development, or development of materials." (GIMMS Spain Coordinator 23/09/2009)

Teachers do not appear to connect with a feeling of shared responsibility to a wider education community. Educational innovation and change in Spain is difficult to organize officially. GIMMS Spain introduced innovation as collaborative mentoring between schools and the university in the area of materials science. There were innovative teachers in the case study schools doing many interesting things in collaboration with university research groups. However this is not the norm as teachers generally are not that interested in educational innovation and change.

Four schools participated in GIMMS *Spain* examining collaborative and reflective mentoring models between experienced teachers, student teachers and teacher educators. They focused on learning about scientific competencies in context-oriented science. Mentoring is not generally 'problematised' with a perception that a good mentor automatically follows from being a 'good teacher':

> "Mentors are not educated, in Spain they are just teachers, expert teachers, with no (education) and training and its not easy to speak about mentoring, because mentoring is supposed to be natural, if you are a good teacher then you are meant to be a good mentor". (GIMMS Spain Coordinator 23/09/2009)

There was some measure of success for GIMMS Spain. Teachers became involved in a deliberative discourse with teacher educators and researchers. While at the start of the project mentoring was largely conceived as a novice-expert relationship it was increasingly conceived within a collaborative and co-inquiry framework toward the end. These changed conceptions are continuing to be developed in other projects, such as COMPEC, that have since started. GIMMS Spain teachers were equally reluctant, alongside teachers in Ireland and the Czech Republic, to include gender as a focus within the project. Experienced science teachers argued that focusing directly on gender would run the risk of stereotyping students. Taking context into account was perceived as a sufficient approach to including both girls and boys:

> "When you work on higher-order thinking and teach science for understanding and … use argumentation and language and strong context for each activity, girls feel that they can contribute more to these tasks, so we were not working on a gender project but we use what we know about gender to produce activities that were gender sensitive in this way". (GIMMS Spain Coordinator 23/09/2009)

While GIMMS Spain made inroads in developing a co-inquiry model of mentoring, mutual learning of this nature was only reported by three mentors out of thirteen mentors in the case study. The discursive 'struggle' was evident in the 'difficulty' reported by experienced teachers when challenged with exploring new ways of co-planning and reflecting on their science teaching. What was successful in the project was the willingness of experienced teachers and student teachers to prepare new web-based resources in materials science.

Teacher education in *Germany*, as a result of PISA, became involved in a central effort to develop innovative processes. The sixteen federal states are very different in learning outcomes and educational policies. Teacher education is organized in two stages: in the first stage at the university (pre-service) students mainly gain theoretical knowledge. In a consecutive second stage beginning teachers get involved in more practical experiences at school. Student teachers gain some support for their lesson preparation in cooperation with mentors or supervisors. More extended cooperation in teams is the exception. The nationwide project SINUS (Prenzel, 2000), following on from the PISA results, tries to improve in-service teacher education with more cooperation in school settings. Both stages of teacher education were represented by the two different projects: GIMMS *Germany* I and GIMMS *Germany* II.

GIMMS *Germany* I involved generating innovation in a framework that discussed context oriented biology education. It was developed as a course for biology student teachers about ethical issues in a number of topics in biology. New and innovative gender sensitive classroom materials were developed and reflected upon. The processes that led to these resource materials was the type of 'deliberative discourse' generated by experienced teachers, student teachers and the teacher educator. The underpinning principles were those of a collaborative, reflective and dialogical mentoring with student teachers on a number of different levels, including personal professional development. This expansive model of mentoring, adopted from Niggli (2003, 2004), was judged to be a successful way to change mentoring relationships from top down hierarchical expert-novice approaches to a model characterized by flat structures, mutual respect and dialogic exchange. The project initiated a deliberative discourse about values in teacher education and about communication among student teachers, experienced mentor teachers and teacher educators as planning partners. This deliberative discourse also expanded across borders between different countries, Germany and Austria. Materials were disseminated through a number of national and international conferences.

GIMMS *Germany* II was a comparative study of different teacher education approaches in Germany. The sixteen federal states are very different in learning outcomes and educational policies. GIMMS Germany II assessed and analyzed teacher professional development in the second phase of teacher education, during their school-based placement experience, in two different states. A central question about innovation in this study was how to develop better partnerships between beginning teachers and experienced teachers for continuing professional development. Teacher education in the federal state Schleswig-Holstein supports autonomy and critical feedback for teachers to a certain extent. This was different from a more conservative model of study seminars, found in the other state Niedersachsen that was used for comparison purposes. Strengthening responsibilities in schools, for teacher continuing education and mentoring, was considered an important outcome of this comparative analysis. This supported a deliberative discourse of different stakeholders based on more practical needs and requirements for teaching. However it also produced some confusion for beginning teachers as they felt forced to follow different masters in school and state institutes. "The care of students has become broader through the new model and is differentiated through the new education of mentors" (Interview GF). This new approach required the strengthening of responsibilities in schools and supported a discourse of different stakeholders in the school setting with more practical requirements for teaching.

In *Austria* PISA results initiated various structural changes in the school system and innovative projects in educational teacher research. The Ministry of Education started to centralize the education system and introduce testing. In addition the system was seeking to recruit more girls into science and technology related subjects. Generally in Austria schools have a lot of autonomy. There are many examples of 'thinking teachers' – teachers who are willing to engage in research and write reflective papers, while working alongside teacher educators and in study circles at their schools. This work has been proceeding since the action research movement began in the 1980s by Peter Posch and later facilitated by John Elliott (Posch, 2010, Elliott, 1991).

According to the GIMMS Austrian coordinator reflective teachers make a real difference in their classrooms and schools. Their students notice the level of interest their teachers take in them, what they are learning and how they are learning. This creates a different classroom climate and fosters discussion and feedback about learning. However this approach to teaching is not systematic and it is not what is experienced in all schools by all teachers. Most change and innovation is expected to be generated at the school site and there are fewer examples of teachers crossing boundaries, with teacher educators and others, to extend their professional learning.

In GIMMS Austria a practical approach for physics teaching was designed to support student teachers by developing and discussing lesson concepts. This gave support to student teachers to reflect on their own teaching methods. Student teachers were introduced to a reflective mentoring approach in order to prepare them to think about and research their own teaching. GIMMS Austria supported a 'deeper level of reflection' and assisted in understanding the 'importance of cooperation between teachers and others' (GIMMS Austria Coordinator 29/09/09). Teacher educators used video analysis and research cycles with student teachers to examine gender issues in their teaching. Gender is 'complex' and introduces a messy 'human aspect' alongside the objective physics subject matter. This requires interpretation, often avoided by physics teachers. Student teachers were assisted in this video analysis work through a questionnaire which helped them focus on a number of key questions, for example:

> "What kind of questions did the teacher put, who answers? What do you think? On the content level the physics teacher was saying that the boys are very good in their work and very active. For example, what do you mean by this, what does it mean that they are good and what do you mean by active and why?" (GIMMS Austria Coordinator, p. 11, 29/09/09)

Student teachers got involved in innovative topics such as science labs or ICT which were different from traditional transmission of subject matter knowledge. They were encouraged to learn more about teacher education, mentoring and different teaching methods. Mentoring was developed as a new way of dialogical co-inquiry and reflection in a team that crossed boundaries between student teachers, experienced teachers and teacher educators. Such a team was an essential requirement for a deliberative discourse on traditional teaching of subjects, gender equality and collaboration across borders. Innovation in Austria was interpreted as leading *bottom-up* change within a national top-down reform measure which provided national *framing for the school*. GIMMS Austria gave an opportunity to reflect on 'the different ways mentoring is done in the Austrian context' (GIMMS Austria Coordinator, p. 9-10, 29/09/09). The teacher educator and team worked with student teachers and experienced teachers to develop a democratic framing for mentoring:

> "Reflective models of mentoring, organic with action research models, dialogical, egalitarian models of mentoring, the team works together to develop this (approach)". (GIMMS Austria Coordinator, p. 10, 29/09/09)

In GIMMS *Denmark* case study schools focused on experienced teachers' co-planning with student teachers and developing teachers' higher order cooperation with each other and with teacher educators. In Denmark the experienced teacher is the supervisor of the student teacher during their practice placement. Traditionally teaching is perceived as 'private' and experienced teachers do not generally open their classroom doors for observation or feedback. Pupils' needs are at the center of any discourse between teacher educators, student teachers and experienced teachers. Subject matter knowledge takes less pride of place, in the lower secondary education system. In some instances teachers may not be qualified in subject matter knowledge.

GIMMS *Denmark* focused on raising student motivation and gender sensitive practices through a model of mentoring based on collaborative learning, between teachers, teacher educators and with pupils in the classroom. This involved exploring effective models for the organization of the learning environment – for example, different ways of doing effective group work – and developing gender sensitive materials for teaching for improved motivation. Teacher educators introduced the topic of gender through a seminar with a leading scholar from the University of Copenhagen. They continued this deliberative discourse, from 2006–2009, through selected readings and discussions with teachers, both at face-to-face meetings and through a virtual learning platform. While many student teachers were pursuing gender as a topic for their Bachelor dissertations *GIMMS Denmark* generally encountered high levels of silence and resistance at the level of the staff room:

> "We met with 'gender silence' in the staff, they really didn't think that it was, gender was an issue, we had talked about it, the teachers were saying that gender is not so important, the teacher education college (itself) does nothing (specifically) around gender but yes, around multicultural issues but nothing about gender". (GIMMS Denmark Coordinator 26/09/09)

GIMMS Denmark was successful in many ways. The shape of mentoring changed from traditional supervision to a more collaborative planning approach where student teachers prepared lessons with experienced teachers (mentor teachers), while working alongside teacher educators. It was this co-planning and reciprocal learning between student teachers, experienced teachers and teacher educators that was 'new' and innovative about GIMMS Denmark. As a cultural challenge to the education system it met with limited success. Experienced teachers did engage with the project, some student teachers engaged with it on a voluntary basis for three months. Teacher educators were at the school site for some of this co-planning. Together they generated, in real time and in virtual learning environments, a range of classroom resources for the teaching of nuclear energy and other science topics. Their deliberative discourse was successful in challenging existing supervision roles:

> "(The) definition of our role is not the traditional supervisor role, it's a more collaborative version of mentorship, they are preparing the teaching together (student teacher and experienced teacher) and they are more in eye-sight with (us), they are coming from different positions and they merge that

knowledge and every side is having new knowledge". (GIMMS Denmark Coordinator 26/09/09)

GIMMS Denmark coordinators considered that their findings illuminate the 'problems and possibilities' of teacher educators working in collaboration with experienced teachers. The possibilities included 'newer' ways of co-planning, mutual learning and reflection between student teachers, experienced teachers and teacher educators. This generated better connectivity between research findings and classroom practices. GIMMS Denmark offered a way of strengthening the stance of the teacher as a reflective practitioner. Currently this stance is not fully embedded in teaching in Denmark:

> "They are expected to have this (reflective awareness) but they don't have it, it is part of the professional code, we are a long way from that but are moving in a new direction with schools ... a lot of teachers who are educated in the last ten years are prepared for it, but the teachers who are educated before that, they are not prepared for it ... it will be (easier) for the student (teacher) to include theory and practice, we have (had) such difficulties in introducing theory to the experienced professional teachers". (GIMMS Denmark Coordinator 26/09/09)

Pupils involved in GIMMS Denmark found the 'new' approaches to teaching motivating and innovative. The collaborative discourse across different teacher communities, while it had many successes it also had constraints. Student teachers who voluntarily engaged over a three month period with the project did not come to a shared meaning making with other participants. Gender awareness within the curriculum process, while considered important by student teachers and teacher educators, was not considered important by experienced teachers. Teachers and student teachers participated in a Curriculum Workshop with the intention of a deliberative discourse, which was positively appreciated but not conducted taking account of the ethical protocols of equal rights for all participants. The need for 'designated thinking time' to be officially provided for this type of deliberative discourse was raised as an important policy issue of concern into the future.

Cross case comparison about the realization of a deliberative discourse

Towards the beginning of GIMMS a comparative study was conducted using ratings on ten scales about factors supporting a discursive approach in national case studies (Table 4). The ratings recorded were self-ratings controlled by ratings of the other project members (Table 5). These discourse components represented the paradigm shifts taking place across Europe at the time: moving policy from official institutions to more school-based decision structures, from teaching for the test to teaching for inquiry and inclusion and professional development for teachers through reflective forms of mentoring and networking through border crossing with others, including teacher educators and political stakeholders.

Table 4: GIMMS criteria as items of interest to be rated.

Number	Description of Criterion
1	School based planning The influence of planning is shifting from higher education and official institutions to the schools.
2	School development Changes in school are intended to recreate the teaching profession for supporting societal change instead of reacting to reforms.
3	Mentoring within teacher education A system guiding and supporting the direct practical experience of teachers about pupils, situations, subject matter and strategies in schools. The key question appears to be: is it for socialisation or for innovation and change?
4	Professional development Hargreaves & Fullan (2000) define four stages of professional development: pre-professional mass education, autonomous professional, collegial professional and post-modern professional teaching.
5	Innovation For example, the introduction of new topics, new methods, new teaching strategies for collaborative learning, new ways of teacher learning.
6	Gender aspects in science and mathematics teaching Through the gender gap girls in schools and women in science and mathematics occupations are underrepresented. For example, girls achieve less in science performances than boys (TIMSS).
7	Gender aspects in teacher education There is a need of more gender awareness among teachers in the way they organise and interact within this learning environment.
8	Networks and ICT: Teacher education with modules for blended learning, simulation and web-pages.
9	Collaboration across boundaries Expanding school activities with a number of different educational and political communities that move teachers beyond narrow school boundaries.
10	Educational policy Collaboration with political and educational stakeholders and policymakers.

Table 5: Judgements from the GIMMS national case studies.

Frequencies	IE	DE I	DE II	DK	CZ	ES	sum
1. School-based TE	1+4	1+1	1+2	1+5	1+3		20
2. School development	0+2	0+1					3
3. Mentoring	1+3	1+4	1+4	1+3	1+3		22
4. Professional development	1+3	1+3	1+3	1+5	0+3		21
5. Innovation	1+2	1+4		1+4	1+4		18
6. Gender aspects in TE	0+2	1+1		0+2	0+2		8
7. Gender aspects in teaching	1+4	1+4	1+1	1+3	1+2		19
8. Net pages / ICT	0+3			1+1	1+3		9
9. Collaboration across boundaries	1+2	1+0	0+1	1+3	0+1		10
10. Policy involvement	1+4	0+1					6

In Table 5 the frequencies with the + signify the self-rating before the + and the rating of the other members in the co-ordination team after the +. Frequencies are on the average and in each case study high for the categories 1, 3, 4 (school-based teacher education, mentoring, professional development), and low for 2 and 10 (school development and policy involvement). Gender aspects are more often mentioned for teaching and less for teacher education. Collaboration across boundaries, networking and policy involvement (categories 8, 9 and 10) were rated with low frequencies. This means that a political discourse was found rather seldom in the national projects, only to a higher degree in the Irish and Danish project.

When we compare these initial results, in the early days of the project, with the former section about deliberative discourse activities for innovation, in the final case study reports from part 2, we find a reasonable increase in all categories. At the end of the GIMMS project it seems, that stakeholders of different communities were engaged in a deliberating discourse for the development of educational innovation and change in science and mathematics teaching and teacher education that crossed borders and was collaborative and democratic.

Mentoring changed during the lifetime of the project to a more collaborative model where mentors, teacher educators and researchers worked together or student teachers prepared lessons with mentors and learned from one another.

In the Danish case the co-planning and reciprocal learning between student teachers and experienced teachers was 'new' and innovative about the project. Experienced teachers and student teachers engaged with it and the teacher educators were at the school site for some of this co-planning. Traditional supervision roles of teacher educators and experienced teachers were challenged in this discourse.

In the Spanish case study with an expert-novice model of mentoring at the outset this kind of co-planning between mentors and student teachers was not of primary interest but seen as a consequence of co-learning between mentors and teacher educators/researchers:

> "Very few mentor teachers recognized at any deep level their role in the development of awareness of sharing of experience in designing teaching activities for teaching 'using scientific evidence' in their classroom. ... In GIMMS Spain the experienced mentor teachers were more open to the co-learning opportunities found within the open 'public space' for discussion with other mentor teachers and the teacher educators/researchers". (GIMMS Spain Coordinator 26/09/09)

Co-planning of mentors and student teachers was seen as important but insufficiently realized in the Germany II case study due to a newly introduced teacher education system. Here the teacher education institute introduced modules for beginning teachers that were not sufficiently tuned with practical work of the schools and mentors to merge their knowledge as in the Danish case. This is critically reflected in the following interview (Mooney Simmie, G. & Lang, M., 2012; Germany II):

> "In general mentors are not well trained. This is obvious ... we as beginning teachers experience, that we are trained very well for the module and then we give our knowledge to the mentors ... They don't know much about it and in our meetings for consultation the mentors are those, who can say the least about the given lesson. There should be done more so that trainers become trainers, which are now simple teachers who guide us."

In this case study a culture of cooperation and autonomy seems to be fostered through partnership based mentoring. But the opportunities for beginning teachers to discuss their work with mentor teachers in accordance with the teaching modules from external teacher educators were estimated to be low. Nonetheless this lower-order co-operation on the practical and technical aspects of professional work that seemed to dominate the mentoring processes was critically reflected for the development of a more inclusive view.

Overall the GIMMS coordinators considered that their findings illuminated the 'problems and possibilities' of teacher educators working in collaboration with experienced professional teachers at the school site. The possibilities included the 'newer' ways that student teachers, experienced teachers and teacher educators could work for co-planning and mutual learning.

This generated the possibility of more collaborative forms of mentorship and better connectivity between the results from research and classroom practices and knowledge generation at the school site. The teacher educators saw this project as a way of strengthening the stance of the teacher as researcher, public intellectual and activist professional working across borders, for the development of a transformative and innovative science education experience for all pupils.

Discussion and implications for the future

GIMMS 2006–2009 was concerned with improving science and mathematics teaching and teacher education. It worked with teachers as extended professionals in a deliberative discourse to generate educational innovation and change. GIMMS concerned itself, in a variety of different ways, with developing the thinking teacher and public educator, as argued by Giroux (1988) to engage in communication, justification and re-conceptualization of teaching with student teachers, with teacher educators and sometimes with policymakers.

GIMMS was set within an evolving theoretical framework that regarded educational innovation and change as political text (Aronowitz & Giroux, 1991). It argued that innovation is brought about by providing public *incubation spaces* and *productive mentoring relationships* that are egalitarian, reflective and democratic and have the capacity to cross cultural borders between teachers, teacher educators, and policymakers. These mentoring relationships of learning need to be robust enough to foster challenge as well as support (Lynch, 1999; Mooney Simmie & Moles, 2011). The aspects of inquiry that were most pertinent to GIMMS were networking through border crossing, questioning, communication and justification, facilitated by a deliberative discourse (Asay & Orgill, 2012).

The national country reports, in part II, show the enormity of the re-culturing task involved in supporting accomplished teachers develop as teacher leaders and researchers of educational innovation and change. In some instances and increasingly, over the lifetime of the project, this discourse was broadened to a political discourse. This served to breathe life into educational innovation and change generated from the *bottom-up* and involved a new role for policymakers and administrators as they had an inbuilt and agreed freedom to be supportive, neutral or contradictory of national mandates and policy reform measures.

GIMMS indicated a new role for teacher educators, and others, such as policymakers, in the development of educational innovation and change. Agreements between all stakeholders ensured that relationships were based on mentoring as partnerships in innovation that were mutual and based on democratic principles. These collaborations, in a number of diverse settings and with a range of different groupings, provided a public incubation space for the development of the teacher as thinker and agent of change. The study gave 'voice' to the multiple dilemmas and challenges facing science and mathematics teachers on a daily basis in their classroom practice.

At the start of GIMMS experienced teachers across the six European countries were in varying measure working in isolated *private spaces* in their classrooms and schools. Mentoring within each country was largely perceived within a narrow top-down novice-expert framing. GIMMS resulted in tangible changes of practice in both of these arenas. Teachers entered 'public spaces' that crossed borders to discuss and debate educational innovation and change. For example, experienced teachers in GIMMS Denmark became planning partners with student teachers. Experienced teachers in GIMMS Ireland became partners in reflective writing and learning with each other, with teacher educators and with their pupils. Teachers in GIMMS Germany I became planning partners with teacher educators and crossed subject boundaries to engage with ethical issues in the teaching of biology. By the end of the project the complexity and possibility of the mentoring construct had become far more evident.

Mentoring in teacher education used the opportunity to generate reflective collaboration across borders leading to innovative processes. This innovative aspect involves third order political knowledge of teacher educators in addition to the usual requirements of first order localized practitioner knowledge and second order contextual knowledge about teaching adult experienced teachers as Murray and Male suggest (2005, p. 131).

> "We define such mentors as being involved in second-order work (in that they are inducting student teachers into the profession), but it is important to note that this work takes place within the first-order settings of their schools, drawing on their localized, practitioner knowledge of those settings in order to induct student teachers."

Looking ahead we find a significant challenge in the educational system through enhancing teacher education and giving teacher educators support and opportunities to reflect on and analyze their emerging practice. This will not only be a challenge for "second order practitioners" specifically learning about teaching teachers but also for changing the system as "third order practitioners" with expansive learning in public spaces. With Boyd and Murray (2011, p. 34) we find it helpful "to frame the changes identified into the wider framework of the institution and the corporate plan".

Funding within the project, by teacher education institutions in collaboration with a European Commission Comenius project, facilitated the project. It supported the setting aside of 'designated thinking time and planning time' for teacher educators, teachers, and sometimes policymakers, in enabling *public incubation spaces,* for face-to-face and virtual meetings with different groupings within a variety of diverse settings. It also supported the setting up of web-site pages, access to relevant research literature, joint preparation of reflective diaries and the printing and dissemination of a number of resources.

In summary, GIMMS developed innovation through a range of well-defined intellectual, moral, fiscal and logistical supports and structures that provided the 'thinking time', 'border-crossing public structures' and 'support mechanisms' needed for this development to be supported, challenged and facilitated. It presents evidence from six countries that educational innovation and change matters – in the realm of curriculum change, social change and teacher continuing education. The levels of success achieved in moving the innovation and change agenda forward strengthens the argument that finding new ways to teach science and mathematics is not merely a technical competence-based concern. GIMMS findings show that educational innovation and change occur within mutually enriching public spaces that cross borders with teacher educators, and others, that promote the re-conceptualisation of teaching and teacher education. In GIMMS this reconceptualization took place between interested and committed teachers, across their professional lifespan, and equally interested and committed teacher educators and other actors, such as policymakers. It is imperative for innovative teachers to cross boundaries and move beyond their school and classroom. It equally requires teacher educators, and others, such as policymakers, to engage discursively and collaboratively in this new partnership for educational innovation and change:

> "Models of curriculum and teaching need to be developed, ones that, for example, reduce the division between conception and execution and mental and manual labour … . This requires us to communicate with each other in both formal and informal ways. … In the process, educators can educate these groups, at the same time that they (the educators) are being educated themselves. After all, it is somewhat silly to deny the fact that teachers do know things that tend to work in classrooms. In this way, by working in concert with others, the practice of developing our methods and content will also embody the social commitments we articulate". (Apple, 2012, p. 158)

Bibliography

Aikenhead G.S. (1996). Science Education: Border Crossing into the Subculture of Science. *Studies in Science Education* 27, 1-52.

Alsop, R., Fitzsimons, A. and Lennon, K. (2002). *Theorizing Gender*. Cambridge: Polity Press.

Altrichter, H. & Posch, P. (2003). *Lehrer erforschen ihre Unterrichtspraxis* [Teachers investigate their teaching practice], Second Edition.. Innsbruck: Studienverlag.

Altrichter, H. & Posch, P. (2009). Action Research, Professional Development and Systemic Reform. In Noffke, S. & Somekh, B. (Eds.), *The SAGE Handbook of Educational Action Research* (pp. 213-225). London: SAGE Publications.

Andersen, F.B. (2000). *Tegn er noget vi bestemmer: evaluering, kvalitet og udvikling i omegnen af SMTTE-tænkningen*. Århus: Danmarks Lærerhøjskole.

Apple, M.W. (2012). *Education & Power*. Second Edition. New York and Oxon: Routledge Education.

Aronowitz, S. & Giroux, H.A. (1991). *Postmodern Education Politics, Culture and Social Criticism*. Minneapolis and London: University of Minnesota Press.

Asay, L.D. & Orgill, M. (2010). Analysis of essential features of inquiry found in articles published in The Science Teacher, 1998-2007. *Journal of Science Teacher Education*, 21, 57-79.

Balle K. & Mølgård, H., (1997). *Skolens udviklingsplan*. Skoleevaluering. Århus: Danmarks Lærerhøjskole. See: http://kif.emu.dk/public_showdocument.do?inodeid=6002.

Beeth, M.E. (1998). Facilitating Conceptual change Learning: the Need for Teachers to Support metacognition. *Journal of Science Teacher Education*, 9 (1), 49–61.

Bennett, J. (2003). *Teaching and learning science: a guide to recent research and its applications*. London: Continuum.

BLK (Bund-Länder-Kommission für Bildungsplanung und Forschungsförderung) (Ed.) (1997) Gutachten zur Vorbereitung des Programms *"Steigerung der Effizienz des mathematisch-naturwissenschaftlichen Unterrichts"*. Materialien zur Bildungsplanung und zur Forschungsförderung, 60. Bonn: BLK.

Boaler, J. (1997). *Experiencing School Mathematics Teaching Styles Sex and Setting*. Buckingham: The Open University.

Bokeno, R.M. & Gantt, V.W. (2000). Dialogic mentoring. Core relationships for organizational learning. *Management Communication Quarterly*, 14 (2), 237–270.

Bolton, G. (2010). *Reflective Practice Writing & Professional Development*. Third Edition. Thousand Oaks, CA: Sage Publications Inc.

Boyd, H. K. & Murray, J. (2011). *Becoming a Teacher Educator: guidelines for induction*. Bristol: Escalate.

Brookfield, S.D. (1995). *Becoming a Critically Reflective Teacher* San Francisco: Jossey-Bass.

Central Statistics Office (2007). *Second level schools and pupils by year, type of school and statistic*. http://www.cso.ie/px/des/Dialog/Saveshow.asp [accessed 7 May 2011].

ChemEd–Ireland (2009). NCE–MSTL, *Preparing for the Future the Importance of CPD for the Chemistry Teacher,* Poster Abstracts. Power, S. & Mooney Simmie, G. Crossing Boundaries in Science Education, p. 18. Limerick: University of Limerick.

Christensen, B., Glindemann, F.-G. & Riecke-Baulecke, T. (2008). *Der Vorbereitungsdienst in Schleswig-Holstein, zusammenfassender Evaluationsbericht* 2004–2007. Kiel: IQSH.

Cochran-Smith, M. (1999). Learning to teach for social justice. In Griffin, G.A. (Ed.), *Education of Teachers. Ninety Eight Yearbook of the National Society for the Study of Education* (pp. 114–144). Chicago, Il.: University of Chicago Press.

Cochran-Smith, M. (2005). The new teacher education: For better or for worse? *Educational Researcher*, 34 (7), 3–17.

Colwell, S. (1998). Mentoring, socialization and the mentor/protégé relationship *Teaching and Higher Education, 3* (3), 313–324.

Cosgrove, J., Shiel, G., Sofroniou, N., Zastrutzki, S. & Shortt, F. (2005). *Education for Life The achievement of 15-year-olds in Ireland in the Second Cycle of PISA*, Dublin 9: Educational Research Centre http://www.erc.ie.

Couso, D. & Pintó, R. (2009a). *Innovative Mentoring for Innovative Science Education: Teaching of Science for achieving Scientific Competence by Student Teachers and their Mentors*. Oral presentation at ESERA Conference. Istanbul. http://www.esera2009.org/.

Couso, D. & Pintó, R. (2009b). *Y después de PISA, ¿qué? Propuestas para desarrollar la competencia científica en el aula de ciencias de profesores en ejercicio y futuros profesores de ciencias*. Oral presentation at the VIII Congreso Internacional de Investigación en Didactica de las Ciencias. Barcelona. http://ensciencias.uab.es/congreso2009/cast/index.html.

Crossley, M. & Watson, K. (2003). *Comparative and International Education Globalisation, context and difference.* NY: Routledge.

Darling-Hammond, L. & Bransford, J. (Eds.) (2005). *Preparing Teachers for a Changing World: What Teachers Should Learn and Be Able to Do.* San Francisco: Jossey-Bass.

Department of Education and Skills (2003). *Junior Certificate Science Syllabus.* Dublin: The Stationery Office.

Department of Education and Skills (2003). *Junior Certificate Mathematics Syllabus.* Dublin: The Stationery Office.

Department of Education and Skills (2008). *Looking at Junior Cycle Science Teaching and Learning in Post-Primary Schools Promoting the Quality of Learning.* Dublin: Department of Education and Skills.

Department of Education and Skills (2011–2020). *Literacy and Numeracy for Learning and Life. The National Strategy to Improve Literacy and Numeracy among Children and Young People.* Dublin: Department of Education and Skills.

Department of Education and Skills web-site for junior cycle syllabuses in science and mathematics http://www.education.ie/juniorcycle.

Dept. Educació. (2007). *Currículum educació secundària obligatòria.* Catalunya. Departament d'Educació 373.5 (467.1).

Driver, R. (1983). *The pupil as scientist?* Milton Keynes: Open University Press.

Driver, R. & Bell, B. (1986). Students' thinking and the learning of science: a constructivist view. *School Science Review, 67* (240), 443–56.

Driver, R. & Oldham, V. (1986). A Constructivist Approach to Curriculum Development in Science. *Studies in Science Education, 13,* 105–122.

Dyer, J., Lindsay, J., Gregersen, H. & Garber, C. (2009). *How do Innovators Think.* Harvard Business Review. Available at: http://blogs.harvardbusiness.org/hbr/hbreditors/2009/09/how_do_innovators_think.htm.

Ecker, A. (2005). *Fachdidaktik im Aufbruch. Zur Situation der Lehramtsstudien an der Universität Wien,* (Subject didactics in break-up. About the situation of teacher education courses at the University of Vienna). Frankfurt am Main: Peter Lang.

Egelund, N. (2007). *PISA 2006 undersøgelsen – en sammenfatning, Danmarks Pædagogiske Universitetsforlag, Danmarks Pædagogiske Universitetsskole,* www.forlag.dpu.dk. Available at: http://www.dpu.dk/site.aspx?p=9707 [accessed 20 September 2010].

Elliott, J. (1991). *Action Research for Educational Change.* Buckingham: Open University Press.

Elster, D. (2006). Interdisciplinary Environmental Education within the training of Teachers for Biology and Physics. In Kyburz-Graber, R. (Ed.), *Reflective Practice in Teacher Education. Learning from Case Studies of Environmental Education* (pp. 101–111). Bern: Peter Lang.

Elster, D. (2007). Contexts of Interest in the View of Students – Results of the German and Austrian ROSE Survey *Journal of Biological Education, 42* (1), 1–9.

Elster, D., Albrecht, R., Hitzenberger, R., Kartusch, R., & Stadler, H. (2006). Praxisorientierte interdisziplinäre Umweltbildung in der Ausbildung von Biologie- und Physiklehrerinnen. In Ecker, A. (Ed.), *Fachdidaktik im Aufbruch. Zur Situation der Lehramtsstudien an der Universität Wien* (pp. 189 – 223). Frankfurt am Main: Peter Lang.

Engeström, Y. (2000). Activity theory as a framework for analyzing and redesigning work. *Ergonomics, 43* (7), 960–974. http://www2.sims.berkeley.edu/academics/courses/is290-3/s05/papers/Activity_theory_as_a_framework_for_analyzing_and_redesigning_work.pdf.

Eurydice (2009). *National summary sheets on education systems in Europe and ongoing reform*. Available at: www.eurydice.org/ [accessed 12 May 2012].

Fensham, P. J. (2007). *Competences, from within and without: new challenges and possibilities for scientific literacy*. Proceedings of the Promoting Scientific Literacy. Available at: http://www.fysik.uu.se/didaktik/lsl/Web%20Proceedings.pdf.

Fischer, D., v. Andel, L., Cain, T., Zarkovic–Alesic, B. & v. Lakerfeld, J. (2008) *Improving School-based Mentoring. A handbook for mentor trainers*. Münster/New York/München/Berlin: Waxmann Publishers.

Freudenthal, H. (1984). *Didactical Phenomenology of Mathematical Structures*. Dordrecht, The Netherlands: Reidel.

GIMMS – *Gender, Innovation and Mentoring in Science and Mathematics* – an EU Comenius 2.1 project, 2006–2009. Available at: http://www.gimms.eu/ [accesssed 12 December 2011].

GIMMS Germany (2009). Available at: http://www.gimms.ipn.uni–kiel.de/ [accesssed 12 December 2011].

Giroux, H. (1988). *Teachers as Intellectuals Toward a Critical Pedagogy of Learning*. USA: Bergin and Garvey.

Giroux, H. (2009). *Against the Terror of Neo-Liberalism*. London: Paradigm Press.

Giroux, H.A. & McLaren, P. (1986). Education and the Politics of Engagement: the Case of Democratic Schooling. *Harvard Educational Review, 56*, 3.

Gott, R., Duggan, S. & Roberts, R. *Concepts of Evidence*. University of Durham. Available at: http://www.dur.ac.uk/richard.gott /Evidence/cofev.htm.

Habermas, J. (1984). *Vorstudien und Ergänzungen zur Theorie des kommunikativen Handelns*. Frankfurt am Main: Suhrkamp.

Habiballa, H., Volná, E. & Fojtík, R. (2006). Interdisciplinarity in teaching numerical analysis. In *4th Mathematical Conference* (pp. 85–93). Nitra: Faculty of Natural Sciences, Constantine the Philosopher University.

Hammersley, M. (1986). *Case studies in classroom research*. Milton Keynes: Open University Press.

Hansen, K.–H., Gräber, W., & Lang, M. (Eds.) (2012). *Crossing Boundaries in Science Teacher Education*. Münster/New York/München/Berlin: Waxmann Publishers.

Hargreaves, A. & Fullan, M. (2000). Mentoring in the New Millennium. *Theory into Practice, 39*, 50–56. http://www.eduweb.vic.gov.au/edulibrary/public/staffdev/teacher/induction/.

Henningsen, I. (2005). *Et kritisk blik på opgaverne i PISA med særlig vægt på matematik*, MONA. No.1, 2005. The Danish University of Education.

Heuvel–Panhuizen, M. (2006). *Flickproblem och pojkproblem*. In: Boesen, J., Emanuelsson, G., Wallby, A. & Wallby, K. (Eds.), *Lära och undervisa matematik – internationella perspektiv*. Göteborg: Nationellt Centrum för Matematikutbildning.

Hogan, P., Brosnan, A., de Roiste, B., Mac Alister, A., Malone, A., Quirke–Bolt, N. & Smith, G. (2008). *Learning Anew*. Maynooth: National University of Ireland, Maynooth.

James, E., Eikelhof, H., Gaskell, J., Olson, J., Raizen, S. & Saez, M. (1997). Innovation in science, mathematics and technology education. *Journal of Curriculum Studies*, 29 (4) 471–483.

Jaworski, B. (1994). Chapter 22: Being mathematical within a mathematical community. In: Selinger, M. (Ed.), *Teaching Mathematics* (pp. 218–231). Buckingham: The Open University.

Jeong, H., Songer, N. & Lee, S.-Y. (2006). Evidentiary Competence: Sixth Graders' Understanding for Gathering and Interpreting Evidence in Scientific Investigations, *Research in Science Education*, 37, 75–97.

Klieme, E., Artelt, C., Hartig, J., Jude, N., Köller, O., Prenzel, M., Schneider, W. & Stanat, P (Hrsg.) (2010). *PISA 2009, Bilanz nach einem Jahrzehnt*. Münster/New York/München/Berlin: Waxmann Publishers. Zusammenfassung: Available at: http://pisa.dipf.de/de/pisa-2009/ergebnisberichte/PISA_2009_Zusammenfassung.pdf/view [accessed May 2012].

KMK (2009). Kompetenzprofil und Qualifizierungskonzept für Beraterinnen und Berater für Unterrichtsentwicklung. http://www.kmk-format.de/material/beratung/handreichung-201108.pdf [accessed July 2012].

Koballa, T.R., Bradbury, L.U., Glynn, S.M. & Deaton, C.M. (2008)Conceptions of science teacher mentoring and mentoring practice in an alternative certification program. *Journal of Science Teacher Education*, 19, 391–411.

Krainer, K., Dörfler, W., Jungwirth, H., Kühnelt, H., Rauch, F. & Stern, T. (Eds.) (2002). *Lernen im Aufbruch: Mathematik und Naturwissenschaften. Pilotprojekt* IMST². Innsbruck: Studienverlag.

Kühnelt H. & Stadler H. (1997). Combined updating on science and pedagogy for experienced teachers. *Research in Science Education* (RISE), 27(3), 425–444.

Kühnelt, H. & Stadler, H. (2008). How to foster teachers' professional development by action research. In Mikelskis–Seifert, S., Ringelband, U. & Brückmann M. (Eds.), *Four Decades of Research in Science Education – from Curriculum Development to Quality Improvment* (pp. 207–219). Münster/New York/München/Berlin: Waxmann Publishers.

Lang, M. (2006). *How to Improve Science Teaching in Europe: Focusing on teachers' voice in professional development*. In Lang, M., Couso D., Elster, D., Hansen, Mooney Simmie, G. & Szybek, P. (Eds.): *Professional Development and School Improvement. Science Teachers' Voices in School-based Reform*. Innsbruck: Studienverlag.

Lang, M. (2009). *Collaborative Settings to Cross Boundaries in Science Teacher Professional Development: Theoretical implications for Innovation*. ESERA 2009 Istanbul.

Lang, M. (2012). Innovation in the Science Curriculum. The Intersection of School Practice and Research. In Hansen, K.-H., Gräber, W. & Lang, M. (Eds.), *Crossing Boundaries in Science Education* (pp. 33–48). Münster/New York/München/Berlin: Waxmann Publishers.

Lang, M., Couso, D., Elster, D., Klinger, U., Mooney Simmie, G., & Szybek P. (2007) (Eds.) *Professional Development and School Improvement. Science Teachers' Voices in School-based Reform*. Innsbruck: Studienverlag.

Lind, G. (1978). Wie misst man moralisches Urteil? (How can we measure moral decision?). In Portele, G. (Ed.), *Sozialisation und Moral* (Socialisation and Moral). Weinheim: Beltz.

Loughran, J. (2006). *Developing a Pedagogy of Teacher Education: Understanding teaching and learning about teaching*. London: Routledge.

Lynch, K. (1999). *Equality in Education*. Dublin: Gill and Macmillan Ltd.

Lynch, K. & Lodge, A. (2002). *Equality and Power in Schools: Redistribution, Recognition and Representation*. London: Routledge Falmer.

Matthews, M.R. (2000). Constructivism in Science and Mathematics Education. In Phillips, D.C. (Ed.), *National Society for the Study of Education, 99th Yearbook* (pp. 161–192). Chicago, University of Chicago Press. Available at: http://wwwcsi.unian.it/educa/inglese/matthews.html [accessed May 2012].

Maynard, T. & Furlong, J. (1995). Chapter 1: Learning to Teach and Models of Mentoring. In Kerry, T. & Shelton Mayes, A. (Eds.), *Issues in Mentoring* (pp. 10–24). Buckingham: The Open University.

Mayring, P. (1998). *Qualitative Inhaltsanalyse: Grundlagen und Techniken* [Qualitative content analysis: Basics and methods]. Weinheim/Basel: Beltz.

McLaren, P. (1991a). Critical pedagogy: Constructing an arch of social dreaming and a doorway to hope. *Journal of Education, 173* (1), 9–34.

McLaren, P. (1991b). Essay review of 'Literacy: Reading the word and the world' by Freire & Macedo, 1987. In Minami, M. & Kennedy, B.P. (Eds.), *Language Issues in literacy and bilingual/multicultural education*. Cambridge, MA: Harvard Educational Review.

Merriam, S. B. (1998). *Qualitative research and case study applications*. Second Edition. San Francisco: Jossey-Bass Publishers.

Mooney Simmie, G. (2007a). Teacher Design Teams (TDTs) – building capacity for innovation, learning and curriculum implementation in the continuing professional development of in-career teachers. *Irish Educational Studies, 26* (2), 163–176.

Mooney Simmie, G. (2007b). Chapter 4: Mentoring in Science Teacher Education and Professional Development: Exploring new ways in a school-university partnership in the Republic of Ireland. In Lang, M., Couso D., Elster, D., Hansen, Mooney Simmie, G. & Szybek, P. (Eds.), *Professional Development and School Improvement. Science Teachers' Voices in School-based Reform*. Innsbruck: Studienverlag.

Mooney Simmie, G. (2009). *The policy implementation process in the upper secondary education system (senior cycle) and videregående skolen in science and mathematics in the Republic of Ireland and the Kingdom of Norway from 1960-2005*, unpublished PhD thesis. Dublin: Trinity College Dublin.

Mooney Simmie, G. & Lang, M. (2012, in preparation). *Public incubation space for leading educational innovation and change in science teaching and teacher education*.

Mooney Simmie, G. & Power, S. (2012). Chapter 10 Innovations in Science Education Through School University Partnership. In Hansen, K.-H., Gräber, W. & Lang, M. (Eds.), *Crossing Boundaries in Science Education* (pp. 233–254). Münster/New York/München/Berlin: Waxmann Publishers.

Mooney Simmie, G. & Moles, J. (2011). Critical Thinking, Caring and Professional Agency: An Emerging Framework for Productive Mentoring. *Mentoring & Tutoring: Partnership in Learning, 19* (4), 465–482.

Mooney Simmie, G. & Moles, J. (2012). Part Three: Culturally Based Concepts: Chapter 7. Educating the Critically Reflective Mentor. In Fletcher, S. J. & Mullen, C. A. (Eds.), *The Sage Handbook of Mentoring and Coaching in Education* (pp. 107–121). London/California/New Delhi/Signapore: Sage Publications.

Mullen, C. A. (2005). *Mentorship* New York: Peter Lang Primer.

Mullen, C.A. & Lick, D.W (1999). *New directions in mentoring. Creating a culture of synergy*. London: Falmer Press.

Murray, J. & Male, T. (2005). Becoming a teacher educator: evidence from the field. *Teaching and Teacher Education, 21*(2), 125–142. Available at: http://www.sciencedirect.com/science/article/pii/S0742051X04001295 [accessed March 2012].

National Council for Curriculum and Assessment (2011a). *Toward a Framework for Junior Cycle*. Dublin: NCCA.

NCCA (National Council for Curriculum and Assessment) (2011b). *Innovation and Identity: Ideas for a new Junior Cycle*. Dublin: NCCA.

NCCA (National Council for Curriculum and Assessment) (2011c). *Teacher Guidelines*. Available at: http://www.ncca.ie [accessed May 2012].

Ní Ríordáin, M. & Hannigan, A. (2009). *Out–of–Field Teaching in Post-Primary Mathematics Education: An Analysis of the Irish Context*. A Research Report. National Centre for Excellence in Mathematics and Science Teaching and Learning, Limerick: University of Limerick.

Niggli, A. (2003). Handlungsbezogenes 3-Ebenen-Mentoring für die Ausbildung von Lernprozessen. *Journal für Lehrerinnen- und Lehrerbildung, 3* (4), 8–15.

Niggli, A. (2004). *Standard-based three-level-mentoring in teacher education* Paper presented at the 29th annual ATEE conference 2004 in Agrigento, Italy. Available at: http://www.teml.at/sites/sites/Niggli-3-Ebenen-Mentoring-Artikel.pdf [accessed May 2012].

Nixon, J. (2004). Chapter 7: Learning the Language of Deliberative Democracy. In Walker, M. & Nixon, J. (Eds.), *Reclaiming Universities from a Runaway World* (pp. 114–127). Maidenhead: Open University Press.

Nowotny, H., Scott, P. & Gibbons, M. (2001). *Re-Thinking Science: Knowledge and the Public in an Age of Uncertainty*. London: Polity Press.

OECD (Organisation for Economic and Co-operation and Development) (2001). *Knowledge and Skills for Life. First Results from PISA 2000*. Paris: OECD.

OECD (Organisation for Economic and Co-operation and Development) (2003). T*eachers Matter Background Country Report for Ireland*. Paris: OECD.

OECD (Organisation for Economic and Co-operation and Development) (2004). *Learning for Tomorrow's World. First Results from PISA 2003*. Paris: OECD.

OECD (Organisation for Economic and Co-operation and Development) (2005). *Teachers Matter: Attracting, Developing and Retaining Effective Teachers*. Paris: OECD.

OECD (Organisation for Economic and Co-operation and Development) (2005). *La definición y selección de competencias clave. Resumen ejecutivo del proyecto DeSeCo*. Paris: OECD.

OECD (Organisation for Economic and Co-operation and Development) (2007). PISA 2006 *Science Competencies for Tomorrow's World*. Paris: OECD.

OECD (Organisation for Economic and Co-operation and Development) (2009). *Equally prepared for life? How 15-year-old boys and girls perform in school*. Paris: OECD. Available at: http://www.oecd.org/dataoecd/59/50/42843625.pdf [accessed May 2012].

OECD (Organisation for Economic and Co-operation and Development) (2009). *Teaching and Learning International Study* TALIS *Background Country Report for Ireland* Paris: OECD.

OECD (Organisation for Economic and Co-operation and Development) (2010). *PISA 2009 at a Glance*. Paris: OECD. Available at: http://dx.doi.org/10.1787/9789264095298-en [accessed May 2012].

OECD (Organisation for Economic and Co-operation and Development) *Programme for International Student Achievement* PISA 2000, 2003, 2006, 2009. Paris: OECD. Available at: http://www.oecd.org/statisticsdata/0,3381,en_2649_35845621_1_119656_1_1_1,00.html [accessed May 2012].

Olson J. (2002). Systemic change/teacher tradition: legends of reform continue. *Journal of Curriculum Studies, 34* (2), 129–137.

Osborne, J. (2007). *Engaging young people in Science: thoughts about future directions of science education*. Proceedings of the Promoting Scientific Literacy Symposium. Uppsala. Available at: http://www.fysik.uu.se/didaktik/lsl/Web%20Proceedings.pdf.

Osborne, J., Erduran, S., & Simon, S. (2004). Enhancing the quality of argumentation in school science. *Journal of Research in Science Teaching, 41*, 994–1020.

Perrenoud, P. (2008). Construir las competencias, ¿es darle la espalda a los saberes?, *Revista de Docencia Universitaria*, II (8).

Petty, G. (2009). *Evidence-Based Teaching A Practical Approach*. Second Edition. Cheltenham: Nelson Thornes.
Pinar, W. F., Reynolds, W.M., Slattery, P. & Taubman, P. M. (2008a). Chapter 5: Understanding Curriculum as Political Text in *Understanding Curriculum An Introduction to the Study of Historical and Contemporary Curriculum Discourses* (pp. 243-314). New York etc.: Peter Lang.
Pinar, W. F., Reynolds, W.M., Slattery, P. & Taubman, P. M. (2008b). Chapter 7: Understanding Curriculum as Gender Text in *Understanding Curriculum An Introduction to the Study of Historical and Contemporary Curriculum Discourses* (pp. 358-403). New York etc: Peter Lang.
Pintó, R. & El Boudamoussi, S. (2009). Scientific Processes in PISA Tests Observed for Science Teachers. *International Journal of Science Education, 31* (16), 2137-2159.
Plauborg, H. C. M. (2007). *Aktionslæring – læring i og af praksis*, Hans Reitzels Forlag.
Posch, P. (2010). Teacher Education in Austria. In Karras, K. G. & Wolhuter, C. C. (Eds.), *International Handbook of Teacher Education World-Wide – Issues and Challenges* (pp. 61-82). Athens: Atrapos Editions.
Physical Sciences Journal (2009). For Second Level Teachers of Physics and Chemistry. *Research into Innovation in Science Teacher Education in the University of Limerick*, p. 29. Dublin: Second Level Support Service in association with the Department of Education and Skills.
Prenzel, M. (2000). Steigerung der Effizienz des mathematisch-naturwissenschaftlichen Unterrichts: Ein Modellversuchsprogramm von Bund und Ländern. *Unterrichtswissenschaft, 28,* 103-126.
Prenzel, M., Artelt, C., Baumert, J., Blum, W., Hammann, M., Klieme, E. & Pekrun, R. (Eds.) (2007). *PISA 2006. Die Ergebnisse der dritten internationalen Vergleichsstudie*. Münster/New York/München/Berlin: Waxmann Publishers.
Sachs, J. (2003). *The Activist Teaching Profession*. Maidenhead: Open University Press.
Scantlebury, K. & Baker, D. (2007). Chapter 10: Gender Issues in Science Education Research: Remembering Where the Difference Lies. In Abell, S. K., & Lederman, N. G. (Eds.), *Handbook of Research on Science Education* (pp. 257-286). London: Lawrence Erlbaum Associates.
Schreiner, C. (2006). *Exploring a ROSE-Garden. Norwegian youth's orientations toward science – seen as signs of late modern identities*. Oslo: University of Oslo.
Schreiner, C. & Sjøberg, S. (2006). *The Relevance of Science Education. Sowing the Seed of ROSE*. Oslo: Acta Didactica.
Shockley, K.G., Bond, H. & Rollins, J. (2008). Singing in my own voice: Teachers' journey toward self-knowledge. *Journal of Transformative Education, 6,* 182-200.
Shulman, L.S. (1987). Knowledge and teaching: foundations of the new reform. *Harvard Educational Review*, 1-22.
Sjöberg, S. (2007). Science education and youth's identity construction – two incompatible projects? In Corrigan, D., Dillon, J. & Gunstore, R. (Eds.), *The Re-Emergence of Values in the Science Curriculum*. Rotterdam: Sense Publishers.
Sjøberg, S. (1997). *Science Education Critical Perspectives from Current Research* OECD-Seminar Oslo May 5th to 6th 1997.
Skolestyrelsen (2009). http://www.skolestyrelsen.dk/skolen/afsluttende%20proever/2%20fagene.aspx.
Sørensen, H. & Andersen, A. M. (2007). *Elevers holdninger til og interesse for naturfag og naturvidenskab*, In: Egelund, N. (ed.) *PISA 2006 – Danske unge i en international sammenligning*. Danmarks Pædagogiske Universitetsforlag, Danmarks Pædagogiske Universitetsskole, Aarhus Universitet, www.forlag.dpu.dk. Available at: http://www.dpu.dk/site.aspx?p=9707 [accessed 20 September 2010].

Sørensen, H. (2008). *Piger og drenge svarer forskelligt – hvilke konsekvenser har det for undervisningen?* In: Troelsen, R.P. & Sølberg, J. (Eds.), *Den Danske ROSE–undersøgelse – en antologi, Institut for Curriculumforskning.* Danmarks Pædagogiske Universitetsskole, Aarhus Universitet Marts 2008. Available at: http://www.dpu.dk/site.aspx?p=5865 [accessed 20 September 2010].

Stadler, H., Benke, G. & Duit, R. (2001). How do Boys and Girls use Language in Physics Classes? In Behrendt H. et al. (Eds.), *Research in Science Education – Past, Present and Future* (pp. 283–288). Kluwer Academic Publishers.

Stadler, H. (2003). *Videos as a tool to foster the professional development of science teachers.* Paper presented at the 4th conference of the European Association of Science Education (ESERA), 2003, Noordwijkerhout.

Stadler, H. (2005). Intervention durch Forschung. Wege zur Unterstützung der Professionalisierung von Lehrkräften mittels Video [Research supported intervention. Support of teacher professional development by video]. In Welzel, M. & Stadler, H. (Eds.), *Nimm' doch mal die Kamera! Nutzung von Videos in der Lehrerbildung. Beispiele und Empfehlungen aus den Naturwissenschaften* (pp. 177–196). Münster/New York/München/Berlin: Waxmann Publishers.

Stenhouse, L. (1975). *An introduction to curriculum research and development.* London: Heinemann.

Teaching Council (2007). *Teaching Council Codes of Professional Conduct.* Maynooth: The Teaching Council. Available at: http://www.teachingcouncil.ie [accessed May 2012].

Teaching Council (2011). *Policy on the Continuum of Teacher Education. June* 2011. Maynooth: The Teaching Council.

Undervisningsministeriet (2009a). Available at: http://www.uvm.dk/Uddannelse/Folkeskolen/Fag%20proever%20og%20evaluering/Afsluttende%20proever.aspx [accessed 20 September 2010].

Undervisningsministeriet (2009b). Available at: http://www.uvm.dk/Uddannelse/Folkeskolen/Fag%20proever%20og%20evaluering/Faelles%20Maal%202009.aspx [accessed 20 September 2010].

Varela, M.P. & Pérez de Landazábal, M. C. (2009). Pisa 2006. Competencia Científica para el mundo del mañana. In *Procceedings of the "19º Encuentro Ibérico de la Enseñanza de la Física", Sept. 2009. Ciudad Real (Spain)* (pp. 100–101). Available at: http://bienalfisica09.uclm.es/libroElectronico/pdf/encuentro_iberico.pdf [accessed May 2012].

Wang, J. & Odell, S. J. (2007). An alternative conception of mentor–novice relationships: Learning to teach in reform-minded ways as a context, *Teaching and Teacher Education*, 23, 473–489.

Wenger, E., McDermott, R., & Snyder, W.M. (2002). *Cultivating Communities of Practice*, New York: HBS Press.

Wilson, J. T. & Chalmers, I. (1988). Reading Strategies for Improving Student Work in the Chem Lab., *Journal of Chemical Education*, 65 (11), 996–999.

Yin, R.K. (2003). *Case Study Research Design and Methods*, Third Edition, Sage Publications.

Zohar, A. (2003). Her physics, his physics: gender issues in Israeli advanced placement physics classes, *International Journal of Science Education*, 25 (2), 245–268.

Links:
Navimat, http://www.navimat.dk/ (accessed 20 September 2010)
NTS, http://www.nts-centeret.dk/ (accessed 20 September 2010)
Viten, http://www.viten.no/ (accessed 20 September 2010)

List of authors

Digna Couso, Universitat Autònoma de Barcelona, Spain
    digna.couso@uab.es

Doris Elster, University of Bremen, Didactics of Biology, Bremen, Germany
    elsterd7@uni-bremen.de

Rostislav Fojtík, University of Ostrava, Czech Republic
    ostislav.Fojtik@osu.cz

Gunnar Friege, University of Hannover, Institute for Didactics of Mathematics and Physics
    friege@idmp.uni-hannover.de

Máire Geoghegan-Quinn, EU Commissioner for Research and Innovation

Hashim Habiballa, University of Ostrava, Czech Republic
    Hashim.Habiballa@osu.cz

Manfred Lang, IPN, Leibnitz Institute for Science and Mathematics Education, Kiel, Germany
    langmanfred@googlemail.com

Geraldine Mooney Simmie, University of Limerick, Faculty of Education and Health Sciences, Ireland
    Geraldine.Mooney.Simmie@ul.ie

Susanne Neumann, University of Vienna, Department of Physics Education,
    SusanneNeumann@univie.ac.at

Roser Pintó, Universitat Autònoma de Barcelona, Spain
    roser.pinto@uab.es

Sancha Power, University of Limerick, Faculty of Education and Health Sciences, Ireland
    Sancha.Power@ul.ie

Lotte Skinnebach, University College Lillebaelt, Department of Teacher Education in Jelling, Denmark
    lihs@ucl.dk

Helga Stadler, University of Vienna, Department of Physics Education,
    Helga.Stadler@univie.ac.at

Birgitte Stougaard, University College Lillebaelt, Department of Teacher Education in Jelling, Denmark
    bist@ucl.dk

Eva Volná, University of Ostrava, Czech Republic
    Eva.Volna@osu.cz

The Editors

Dr. Geraldine Mooney Simmie lectures in education at the University of Limerick, Ireland. She is Co-Director of a structured PhD programme in education and Director of a Master's in Education (Mentoring) programme. In her role as Academic Coordinator she is involved in building capacity between schools and university to better integrate theory and research findings with professional practice. Her research interest lies in the area of education policy analysis with a particular interest in comparative education policy and policy as it relates to teacher continuing education and mentoring, especially within science and mathematics education. She has published a number of journal articles and book chapters, and contributed to The Sage Handbook of Mentoring and Coaching in Education, 2012. Her doctoral study is a comparative study of upper secondary science and mathematics education in the Republic of Ireland and the Kingdom of Norway. Geraldine coordinated this EU GIMMS project which was managed by the University of Limerick between 2006 and 2009.

Dr. Manfred Lang, now retired, was senior researcher of the Leibniz Institute for Science Education (IPN) at the University of Kiel, Germany from 1971 until 2008. He studied psychology at the universities in Kiel and Hamburg in Germany and Yale in the USA. He coordinated and conducted research in international, European and German studies such as the IEA study "Computers in Education", the OECD study "Science, Mathematics and Technology Education" the European projects FAST-TECNET (Technology, Education and Culture Network), STEDE (Science Teacher Education development in Europe), SOL (Selbstorganisiertes Lernen in der Kernregion Schleswig-Holstein), CROSSNET (Crossing Boundaries in Science Teacher Education) and GIMMS (Gender, Innovation and Mentoring in Mathematics and Science). From each of these projects many presentations and symposia in conferences were offered and articles and books published. He co-edited Crossing Boundaries in Science Teacher Education (2012) published by Münster: Waxmann.

Index

action research 27, 31, 57, 81, 152, 153, 165
apprenticeship 21, 108, 114, 117-118, 124-125, 128, 134, 149
Austria 7, 11-12, 23, 27-28, 67, 72, 75, 80-84, 87-88, 109, 140-144, 148, 152-153, 168
best practice 24, 46
biology 23, 26, 53, 67, 69, 73-74, 128, 151, 160
border crossing 24, 27-30, 49-50, 70, 139, 145-146, 156, 159
boundary 51, 135
case study 7, 21-22, 25-28, 32, 35, 37, 40, 43, 47, 58, 62-63, 65, 120, 123, 125-126, 134, 137, 149-151, 153, 157-158, 166
change 7, 9, 13-14, 17-18, 20-33, 35, 38-39, 41, 43, 45, 47-51, 54-55, 57, 63-64, 66, 68-69, 75, 78-79, 81, 83-84, 89-90, 96, 108, 110, 114-117, 120-125, 130, 134-139, 143-151, 153, 156-157, 159-162, 166, 167
chemistry 23, 37, 53, 59, 64, 96, 98, 128, 145
collaboration 25, 30, 32, 38, 48, 52, 55-56, 61, 63, 65-66, 96-97, 124-125, 132-133, 135-136, 148-150, 153, 155, 158, 160
competencies 9, 26, 90, 126, 130-131, 135-136, 143, 150
constructivism 112, 115
context 7, 13, 19, 26, 29, 31, 35, 37, 47, 54-55, 58-59, 69, 72-73, 75-76, 83, 90-91, 93-95, 107-108, 110, 116, 120, 124-125, 132, 139, 141, 147-148, 150-151, 153, 163, 166, 169
cooperation 10-11, 29, 32, 37-38, 40, 49-50, 58-59, 62, 68, 78-79, 87, 97, 113, 117, 119, 127, 132, 145, 148, 150-151, 153, 158
culture 18, 25, 45, 50-51, 68, 123, 132, 158, 166
Culture 162
curriculum 7, 21, 25-29, 32-33, 38, 40, 50, 54, 59, 61, 63, 68, 75, 77, 79-80, 85-86, 90, 95, 98, 110, 113, 117-118, 123, 127, 131, 144, 146, 148-149, 155, 161, 165-166, 169

Czech 7, 12, 23, 26, 28-29, 109-121, 140-144, 148, 149, 151
deliberation 24, 146
democratic 7, 10, 14, 18, 20-24, 26-27, 30-32, 50-52, 54-57, 62, 67, 75-76, 81, 137, 139, 144, 146, 153, 157, 159
Denmark 7, 12, 23, 28, 31, 33, 52, 53-59, 61-66, 109, 140-144, 148, 153-155, 160
development 7, 11, 14, 17, 19, 22, 25, 27, 32-33, 38, 40-41, 48-51, 54, 56, 58, 63, 66-70, 73, 76, 78-79, 83-84, 87, 94-95, 107-108, 111, 114-115, 118, 120-121, 123, 126, 128, 131, 137, 139-140, 142-144, 146-147, 150-152, 155-159, 161, 165-166, 169
discourse 7-8, 13, 15, 18, 20-21, 23-24, 26, 29, 32, 35, 37, 40, 47-51, 79, 81, 96, 132, 136-137, 139, 143-155, 157-159
dissemination 11, 17, 27, 47, 73-74, 76-77, 161
educational innovation 8, 18, 20-21, 28, 30-32, 35, 41, 48, 50-51, 57, 63, 68-69, 79, 81, 89, 96, 108, 110, 114, 123-124, 134-139, 144-147, 150, 157, 159-161, 166
framework 7, 13, 17-18, 22, 24-25, 37, 39-40, 54, 70, 89, 91, 94-96, 98, 106, 108, 114, 120, 126, 135-136, 137, 139, 143-144, 147, 150-151, 159-160, 164
gender 9, 13, 18, 21-23, 25-29, 31-32, 37, 39-40, 47, 53-59, 61-66, 69, 72-75, 77-78, 81-83, 85, 98, 111-112, 114, 118-120, 123, 128-130, 133-135, 140, 142, 144-145, 148-149, 151, 153-154, 156, 162, 164, 168, 169
Germany 7, 11-12, 23, 25-28, 30, 67-70, 72-73, 75-76, 79, 109, 123-126, 128, 132-133, 136, 139-144, 148, 151-152, 158, 160, 164
ICT 25, 31, 43, 44, 46-47, 49-50, 56, 59, 64, 113, 116-117, 119, 121, 153, 156-157
implementation 13, 26, 38, 43, 95, 117, 136, 143, 166
improvement 42, 114, 136, 143
inclusion 23, 31, 38-39, 49, 54, 66, 80-82, 88, 120, 128-129, 155

inclusive 9, 22-23, 45, 114, 119, 158
innovation 7, 9-10, 13-15, 17-18, 20, 23-32,
    35, 38-40, 42, 47-52, 54-55, 57-58,
    63-66, 68-69, 74-75, 79-81, 89, 95-96,
    98-99, 107-111, 114-116, 119-121,
    123-124, 128-129, 133-137, 139-140,
    144-153, 156-157, 159-161, 166
inquiry 7, 13, 19, 21-23, 26, 30-31, 39-40,
    49, 59-60, 70-71, 79, 90, 96, 110-111,
    114, 120, 124, 126-127, 130, 134, 144,
    149-151, 153, 155, 159, 162
in-service 19, 30-31, 37-38, 53, 57, 66, 81,
    85, 89-91, 94-96, 98, 151
institution 7, 32, 130, 144, 160
integration 43, 126
interactive 23, 111, 117, 120-121
interdisciplinary 75, 116, 163
Ireland 7, 10-12, 15, 23, 25, 28-29, 33, 37,
    39-40, 42, 46-50, 109, 139-144, 148-
    149, 151, 160, 162-164, 166-167
justification 19, 69, 106, 159
mathematics 7, 9, 13, 17, 20-23, 25-26,
    28-29, 31, 37-40, 42-45, 47, 49-50, 52-
    57, 64, 66, 80, 109-114, 116-120, 123,
    128, 140, 142, 145, 147-150, 156-157,
    159-166
meaning 22, 57, 81, 84, 145-148, 155
mentee 25, 27, 57, 69-72, 89, 94-97, 99,
    106-108, 118, 120, 124, 127, 135, 148
mentor 25-28, 31, 52, 57, 62, 65, 68-74,
    76-80, 89, 94-98, 99, 106-108, 118, 120,
    124, 126-132, 134-135, 148-150, 152,
    154, 158, 163-164, 169
mentoring 91, 94-96, 99, 107-108, 114,
    117-118, 120, 158, 162, 165-167
model 20-21, 23, 27-28, 30-31, 40, 54-58,
    63-65, 68-70, 72, 75, 81, 87, 89, 91,
    95-96, 99, 107-108, 114-115, 117-118,
    124-125, 129, 132-134, 142, 149, 151-
    152, 154, 157-158
motivation 13, 28, 31, 41-42, 45-47, 50,
    54-55, 58-59, 66, 77, 85, 111, 113-114,
    120, 154
network 24, 38, 49, 50, 55, 57, 132, 143,
    148
novice–expert 23, 26, 40, 57, 63, 96, 120,
    149, 150, 160
OECD 13, 31, 38, 49, 52-53, 89-90, 93, 98,
    109, 123, 137, 140-144, 148, 167-168
ownership 7, 13, 23, 39, 50, 56

partnership 17, 26, 29-30, 32, 41, 48-49,
    69, 96-97, 109-111, 114-115, 117, 120,
    124, 132, 158, 161, 166
pedagogy 19, 24, 38, 115, 125, 128, 131,
    142-144, 148, 150, 165-166
physics 23, 26-27, 31, 37, 53, 59, 64, 80, 82-
    86, 96, 98, 111-112, 114, 118-119, 128,
    145, 149, 153, 169
PISA 7, 26-27, 31, 38, 52-53, 57-58, 81,
    83, 85, 89-90, 93-94, 98, 109, 120, 137,
    140-141, 143-144, 149, 151-152, 163-
    165, 167-168
policy 7, 28-29, 31, 33, 35, 53, 66, 111, 120,
    123-125, 129, 134-135, 141-143, 145,
    148, 155-157, 159, 166
political text 7, 17-18, 139, 144, 159
practice 9, 19, 22-24, 28-29, 32, 37, 40, 42,
    46, 50, 54-58, 61-66, 69, 72, 81-84, 94,
    96, 99, 108, 110, 112-113, 116-118,
    123-124, 129-131, 134-135, 137, 143,
    148, 154-155, 160-162, 163, 165, 169
pre-service 30, 78, 89, 91, 94-95, 99, 151
profession 143, 156, 160
professional 7, 10, 13-14, 19, 22-23, 25-30,
    32, 35, 37, 40, 47, 49-50, 59, 62, 66-71,
    79, 81-84, 96, 108, 110, 118, 123-124,
    128, 131-133, 135-137, 139, 142-148,
    150-153, 155-159, 161, 165-166, 169
public space 7, 18-19, 23, 69, 81, 108, 137,
    139, 140, 144, 146-148, 158, 160-161,
    166
reflection 17, 26-27, 30-31, 39, 45, 47-48,
    56, 63, 65, 69, 72, 79, 95-96, 98, 114,
    116, 118, 127, 132-133, 136-137, 150,
    153, 155
reflective mentoring 25, 27, 48, 124, 129,
    150, 153
researcher 10, 29, 98, 159
school-university-partnership 7, 24, 26,
    48-49, 69, 110-111, 113-115, 120-121
science 7, 9, 11, 13, 17, 19-23, 25-29, 31,
    33, 37-45, 47-50, 52-56, 58-61, 66,
    72-76, 80-81, 83-87, 89-91, 93, 95-96,
    98-99, 106, 108-120, 123, 140, 142-145,
    147-154, 156-157, 159-163, 165-169
Science teacher 165
scientific literacy 141, 144, 164
setting 14, 24-25, 27, 32, 35, 60, 69, 110,
    118, 123, 135, 137-140, 143-145, 148-
    149, 152, 160

173

Spain 7, 12, 23, 26, 28, 89, 91, 93-99, 106-109, 140-144, 148-151, 158, 169
student 106
student teacher 7, 11, 22-23, 26-28, 30-32, 35, 47-49, 52, 54, 56-58, 61-65, 67, 75, 80-82, 84-87, 89, 94-99, 106, 108, 115, 119, 125-126, 128, 145, 150-155, 157-160
subject 19, 26, 28, 30-31, 38, 40, 44, 47, 53, 55, 59, 66-67, 69, 74-76, 80, 83, 85-87, 91, 110, 112-114, 116, 125-128, 132, 135, 137, 142-143, 146, 149-150, 153-154, 156, 160
teacher education 7, 13, 19, 20, 23-25, 27-28, 30-32, 37, 51, 54-55, 57, 58, 64-69, 74-75, 80-81, 89, 91, 94-95, 97, 108, 118, 123-130, 133-135, 137, 139-140, 142-143, 145, 147-149, 151-154, 156-163, 166-167
teacher educator 7, 10, 13, 17-20, 22-33, 35, 37-40, 42, 46-51, 55-62, 65-66, 68-69, 72, 89, 91, 94, 96-98, 108, 111-112, 118-119, 120-121, 129, 132, 134, 136-140, 143-144, 146-148, 150-161, 166
university 7, 8, 17, 24, 26, 30-32, 39-41, 44, 46, 48-50, 54, 58, 66-69, 72, 74, 75, 77, 80, 83-87, 96, 109-111, 113-115, 119-121, 125, 134, 145, 150-151, 166
video 23, 27, 43, 60-62, 80-84, 86-87, 153, 169
voice 41, 144, 147, 160, 165, 168
web-site 23, 48, 161, 163

Klaus-Henning Hansen,
Wolfgang Gräber, Manfred Lang (Eds.)

Crossing Boundaries in Science Teacher Education

2012, 240 pages, pb, € 29,90
ISBN 978-3-8309-2595-8

This book presents the outcomes of a transnational EU project about innovation in science teacher education. Guiding questions were how teachers, policy makers and teacher educators collaborate in the process of change and how local background projects respond to opportunities for the exchange of experiences and reflection in terms of a common theoretical framework: the idea of boundary crossing. The book is based on a series of local case studies conducted by local coordinators and contracted teachers. The case studies are supplemented by a cross-case analysis of common and distinct features in the projects and an essay about the relationship between boundary crossing, transformative learning and curriculum theory. Main outcomes are suggestions to im-prove school-based reform in and collaboration for science education.

WAXMANN
Münster · New York · München · Berlin

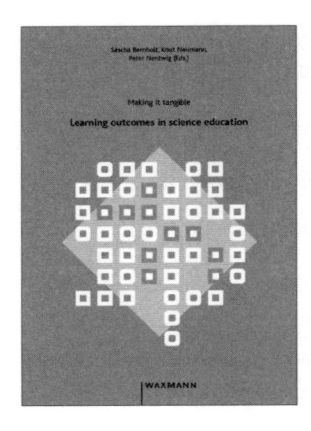

Sascha Bernholt,
Knut Neumann, Peter Nentwig (Eds.)

Making it tangible

Learning outcomes in
science education

2012, 520 pages, pb, € 45,00
ISBN 978-3-8309-2644-3

One of the central features in current educational reforms is a focus on learning outcomes. More recently, the emphasis has shifted to considerations of how standards can be operationalized in order to make the outcomes of educational efforts more tangible. This book is the result of a symposium held in Kiel, and it brings together renowned experts from 12 countries with different notions of the nature and quality of learning outcomes. The aim was to clarify central conceptions and approaches for a better understanding among the international science education community.

WAXMANN
Münster · New York · München · Berlin